Google Pixel 4 /Pixel 4XL User Guide

The Ultimate and Complete Guide to Master the New Google Pixel 4 /4 XL in 3 Hours!

Smith L. Anderson

Copyright 2019 © **Smith L. Anderson**

All rights reserved. This book is copyright and no part of it may be reproduced, stored or transmitted, in any form or means, without the prior written permission of the copyright owner.

Printed in the United States of America.

Copyright 2019 © **Smith L. Anderson**

Contents

Introduction 1

Chapter 1: How to Connect Your Pixel 4 to a Wi-Fi the Way You Want 2

Chapter 2: The Transfer of Data from Your Current Phone 11

Chapter 3: The Reason Why Some Data Doesn't Transfer From an IPhone 19

Chapter 4: Copying Contacts 30

    Importing contacts from a file 30

Chapter 5: Copying Music 37

Chapter 6: Transferring Music from a Computer to Your Phone or Tablet 41

Chapter 7: How to use your Screen 48

Chapter 8: How to Check Your Notifications 51

Chapter 9: Make and Receive Calls on Your Phone 59

    How to Set What To Block 77

Chapter 10: Messages 91

Chapter 11: Choose How You Get Notifications On Your Device . 101

    How to change the notifications for certain apps 101

Chapter 12: How to Set Your Pixel Phone to Automatically Unlock 106

Chapter 13: How to Use the "Ok Google" 120

Chapter 14: Understanding How Your Voice Works With Other Settings 132

Chapter 15: How to Add Apps, Shortcuts, and Widgets to the Home Screen 135

Chapter 16: Home Screens Organization 140

Chapter 17: Managing Screen and the Display Settings 145

Chapter 18: How to Open Quick Settings 157

    Split-Screen Multitasking 167

Chapter 19: How to Take a Screenshot on Your Pixel Phone 173

    Editing Photos 175

    How the Face Grouping Feature Works 182

Chapter 20: Camera and Photos 184

Chapter 21: How to Print from Your Device 197

Chapter 22: Android 10 Navigation Tips 215

    Digital Wellbeing 224

Index 226

Introduction

If you just got for yourself the refreshing Google pixel 4, then we have got some hot tips for you. One of the coolest features about the Google Pixel 4 is the software. The Android 10 update and features that are available on the phone provides you with more control of your device with the new shortcuts, app tweaks as well as other software goodies.

The Pixel 4 is Google's flagship Smartphone. It's what Google says Android should be.

So, in this guide, you are going to learn a lot of great tips that will help get the most from this great device.

Chapter 1: How to Connect Your Pixel 4 to a Wi-Fi the Way You Want

Just like any other phone, your Google Pixel 4 Phone allows you to connect your Wi-Fi the way you want. If you have your Wi-Fi turned on, your phone automatically connects to a close Wi-Fi network you've connected to before.

You can also program your phone to turn on automatically to Wi-Fi near saved networks.

To Connect To A Wi-Fi:

1 Open your device's Settings app.

2. Locate the Network & internet feature and tap on it

3. From there, tap Wi-Fi and Turn it on.

4. Tap a listed network.

5. If it requires a password, you'll see the Lock icon.

After you connect:

You will see "**Connected**" show under the network name. And the network is "**Saved**."

So, when Wi-Fi is on, and your device is near, your Phone will connect automatically.

Tip: The fastest way to get to your Wi-Fi settings is to swipe down on your screen to get to Wi-Fi setting.

Connect To Wi-Fi via Notification

You will receive a notification when your Wi-Fi is on, telling you of available public and high-quality networks.

- To connect to a network when you get these notifications, tap Connect.

But if you don't want to get notifications for that network, just clear the notification.

To set up your phone to automatically connect to open networks:

1. Open your device's Settings app

2. Tap Network & internet and then tap Wi-Fi

3. After that, tap Wi-Fi preferences.

4. Open networks by turning on **Connect.**

3

When Connected Via Wi-Fi Assistant

When your phone is connected to Wi-Fi assistant, your notification bar displays the Wi-Fi assistant virtual private network (VPN) key.

Then Your Wi-Fi connection says: "**Auto-connected to open Wi-Fi.**"

Disconnect or turn off

Your Google Pixel 4 lets you turn off Wi-Fi assistant or disconnect from one network at any time. To disconnect from the current network;

1. Simply open your device's settings.

2. Then tap **Network & internet**, and tap Wi-Fi

3. From there, tap the network name.

4. Tap Forget.

To Turn off Wi-Fi assistant;

1. Open your device's Settings

2. Then, tap **Google** and then **Networking**.

3. Turn off Wi-Fi Assistant.

If you have issues connecting to a nearby open network through Wi-Fi assistant, it could be because:

1. The Pixel interface hasn't verified the network as reliable and high-quality.

2. Wi-Fi assistant will not connect to networks that you've manually connected to.

3. Wi-Fi assistant doesn't connect to networks that require you to follow steps to connect, like signing in.

The solution to these is to:

Connect manually if Wi-Fi assistant does not connect automatically. If you had manually connected to the network already, "**forget**" the network. Wi-Fi assistant will later on re-connect automatically.

Important: Data sent to that network through a manual connection could be seen by other people.

So, to keep your open Wi-Fi network safer, make use of Wi-Fi assistant virtual private network (VPN).

The VPN will help keep your data from being seen by other people sharing the same open network.

If you On a VPN for Wi-Fi assistant, the "**Device connected to Wi-Fi assistant**" message will be displayed.

Also, note that Google uses system data sent via VPN connections to:

1. To offer and improve Wi-Fi assistant, which includes the virtual private network (VPN)

2. To Monitor for abuse

Turn On Automatically Near Saved Networks

You can automatically turn on your Wi-Fi on your Pixel 4 device on near saved networks.

1. Open the Settings app on your device.

2. Then, Tap Network & internet

3. From there, tap on Wi-Fi and Wi-Fi preferences.

4. Then, turn on Wi-Fi automatically.

If needed, you can decide to first turn on **Security & location** and then Advanced. From there, choose Location and then Advanced. After that, select Scanning and then Wi-Fi scanning.

Your Wi-Fi won't turn on automatically when:

1. Battery saver is on

2. Location is off

3. Airplane mode is on

4. Background Wi-Fi scanning is off.

5. Tethering ("hotspot") is on

Add, Change, or Remove Saved Networks

1. Open the settings app on your device.

2. Tap Network & internet and then Wi-Fi.

3. If you want to move between listed networks, tap a network name.

4. Tap a network, and then tap Settings to change a network's settings.

Add Network

To add Wi-Fi networks that don't automatically display in your list.

1. Open the Settings app in your device.

2. Then tap **Network & internet** and then Wi-Fi.

3. Ensure that Wi-Fi is on.

4. Tap Add network at the bottom of the list

If needed, you can fill in the network name (SSID) and security details.

5. Tap Save.

Remove Network

If you do not wish to connect to a saved Wi-Fi network automatically, you can "**forget**" that network.

To have a Wi-Fi network removed from your device:

1. Open the Settings app on your device.

2. Tap on Network & internet and then Wi-Fi.

3. Ensure your Wi-Fi is on.

4. Then Touch and hold a network you've saved.

5. Tap Forget network.

Understanding Proxy Networking

A proxy in your Google Pixel 4 is like a tunnel or gate between devices. For instance, you could be required to connect via a proxy when you connect to a work network from your home.

Connect to a Wi-Fi network via a proxy with these steps:

1 Open the Settings app in your device.

2. Tap Network & internet and then Wi-Fi.

3. Then Tap a network, and tap Settings.

4. Tap Edit at the top.

5. Tap the Down arrow Next to "Advanced options,"

6. Under "Proxy," tap the Down arrow

8. From Down arrow, you can pick the configuration type.

9. If prompted, enter the proxy settings.

10. Tap Save.

You will need to setup proxy settings separately with each Wi-Fi network.

Turn Wi-Fi on near saved network automatically

To do that:

1 Open your Settings app on your device

2. Tap Network & internet and then Wi-Fi.

3. Tap Wi-Fi preferences at the bottom.

4. Tap an option. These vary by Android version and device.

4. Then automatically Turn on Wi-Fi near saved networks.

Other WiFi Features include:

Connect to open networks: You can connect to high-quality open networks automatically.

1 Open network notification and this will give you a notification when automatic connection to high-quality open networks is not available.

Network rating provider: This lets you see ratings for your Wi-Fi networks for your device.

Wi-Fi Direct: This feature allows you to connect your phone without a network with other devices that can utilize Wi-Fi Direct.

Metered Wi-Fi: Set up Metered Wi-Fi to limit how much data your device uses on apps and downloads.

To connect to Metered Wi-Fi;

1 Have your device connected to Wi-Fi.

2. Open your Settings on app device.

3. Tap Network & internet and then Wi-Fi.

4. Then Tap the Wi-Fi network you're connected to.

5. After that, Tap Advanced and then Metered

6. Then Treat as metered.

Chapter 2: The Transfer of Data from Your Current Phone

If you are turning on your Pixel for the first time, you will need to follow the On-screen instructions to get data moved from your Pixel phone.

In a situation whereby you must have skipped data transfer or didn't finish Setup the first time, you will be notified with a message that reads "Pixel setup isn't done."

Once this comes up, tap "**Finish setup**."

Afterward, for a few more days, you will need to take the following steps:

1. Launch your Settings app
2. Tap on Finish Setup at the top.

You can always have your phone reset though. But this will erase all your data.

From an Android phone:

Before you transfer data, it is important for you to take note of the following:

1. It is mandatory to make use of a Nano-SIM card, so if you don't have one, you will need to get in touch with your mobile service provider.

2. Make sure you are using the right cable that works perfectly with your current phone such as the charging cable

3. Ensure that your connection cable is USB-C, otherwise, you should get the Quick Switch Adapter.

Transferring Data from An Android Phone To A Pixel

STEP 1: What can be copied to your Pixel?

- Music, Photos, and Videos

- Apps and app data

- Contacts which are stored on your device or SIM card

- Google Accounts

- Text messages without photos, videos, or music

- Most of the Device settings (depends on the device as well as the Android version)

Whenever you sign in into your Google account using your Pixel phone, you will see the following:

- Contacts

- Email

- Calendar events

- Any other information related to your Google account

- It is to be noted that all contacts, as well as calendars that are copied to your Pixel phone, will automatically sync and get uploaded to your Google account online. When you get signed in to your Google account on your Pixel phone, all information that is associated with your Google account will be synced to it.

What Won't Get Copied During the Setup Process

The following data won't get copied after setup:

1. Photos, videos, and music which are stored in hidden folders.

2. Photos, videos or music present in text messages. Nevertheless, words present in MMS will get transferred but the attachments won't.

3. Downloads such as PDF files.

4. Apps that do not come from the Google Play Store.

5. Accounts different from Google accounts as well as their data.

6. Data that comes from the apps that don't use Android backup.

7. Contacts and calendars that are synced to services that are not from Google.

8. Ringtones

9. Some device settings (depends on the device and Android version)

STEP 2: GET SET TO COPY

1. Make sure both phones are well charged.

2. Get all available updates installed on your current phone.

3. Make sure all you will need for the process is complete by checking what comes with your Pixel phone.

• Check the cable that works with your current phone such as the cable used in charging it.

• In a situation where your current phone or cable has a different connection different from USB-C, then use Quick Switch Adapter.

• If you are not making use of the Google Fi as your Mobile carrier and using an eSIM, then check your SIM card as well as the SIM card insertion tool.

4. Get the following done on your Pixel phone:

a. Get your SIM card inserted: It should be noted that all Pixel phones can make use of the Nano-SIM cards and some Pixel phones also make use of the eSIM.

For users in the United States, you have the option of picking no SIM card or making use of a pre-inserted Verizon SIM card.

In a case where you have the Verizon as your mobile carrier, get your sim activated by visiting www.vzw.com/google-activate.

For users in other countries, the phone won't come with a SIM, so if a SIM is needed. Go get a Nano-SIM card and provide your mobile service provider with your phone's IMEI number of asked.

To locate your phone's IMEI number as earlier said

- Get the Phone box

- Check the phone's SIM card tray.

- Go to your phone's settings: Click on "System" and tap about the phone, after which you should have access to the "IMEI." You are advised to take note of your IMEI number by getting it written down on a paper or taking a picture of it.

b.   Put on your Pixel phone as soon as you are done inserting your SIM.

c.   Normally, you should see a Start button but if not, return to set up.

STEP 3: Copying your data

1.   On the Pixel phone:

a.   Click on Start.

b.   Get the phone connected to a Wi-Fi or a mobile network.

c.   Select "Copy your data."

2. Get your current phone turned on and unlocked.

3.   Get both ends of the cable connected to your current phone and Pixel phone or get it plugged to your Quick Switch Adapter and adapter plugged into your Pixel phone.

4.   Tap Copy on your current phone and follow the onscreen instructions.

5.   A list of your data will appear on your Pixel phone:

- To get all your data copied, tap on Copy.

- To get some of the data copied, turn off what you are not copying and tap copy afterward.

6. You will get notified as soon as the transfer is done.

Troubleshooting Transfer of Data

In a situation where you didn't transfer your data when you turned on your phone for the first time:

a. Within a few minutes, you will be notified with a message that reads "Pixel setup isn't done," tap on "Finish setup," once you see this.

b. For a few days, you will need to take the following steps:

1. Launch the "Settings app ."

2. Tap "Finish setup," at the top.

You can always get to reset your Pixel phone though this will result in deleting all the data on your phone.

c. After you have had your hone a while, you can always get to reset your Pixel phone though this will result in deleting all the data on your phone.

From an iPhone

Before transferring data from an iPhone, you need to take note of the following:

• It is compulsory for you to get a nano-SIM card, so if you don't have one, you should get one from your Mobile service provider.

Nevertheless, some Pixel phones make use of eSIM and this depends on the device or mobile carrier.

• Make sure you get a cable that works perfectly with your current phone like the charging cable

• Make use of the Quick Switch Adapter

Chapter 3: The Reason Why Some Data Doesn't Transfer From an IPhone

Supervised iPhone

Your data will fail to copy if it is being supervised by a business, school, or some other organization. To check if your iPhone is supervised or not, take the following steps:

1. Launch the Settings.

2. Tap on General.

3. Tap "About"

4. Look out for "This iPhone is Supervised," under the name of your device.

Managed iPhone

Copying of data could be limited if your iPhone is being managed by a business, school or other organization as well. To check if your iPhone is managed or not, take the following steps:

1. Launch the Settings.

2. Tap on "General."

3. Tap on "Profiles & Device Management."

4. Look out for Policy profiles from your organization under the device name.

NOTE: Though removing these profiles can help increase what can be transferred, it can affect the access to your organization's information.

Backup ITunes Encryption Turned On

The copying of data will get limited if the backup encryption for your iPhone in the iTunes application on your computer is turned on. To get this turned off, take the following steps:

1. Start iTunes on your computer.

2. Get your iPhone connected to your computer.

3. Navigate to the Summary tab for your iPhone.

4. Get the Encrypt iPhone backup unchecked.

5. Assign a password to the backup encryption.

6. While encryption is off, wait while iTunes does a backup.

7. Get your iPhone disconnected from your Computer.

iMessage Turned On

To ensure that you get all the text messages on your new device, you should turn off the iMessage when switching from an iPhone to a Pixel phone. It is important for you to switch the iMessage off before removing the SIM card from the iPhone to prevent all your messages from going into the old iPhone instead of your new device.

Turning off iMessage

1. Go to the Settings on your iPhone.

2. Tap on Messages.

3. Set the iMessages to off.

4. Get the group chats restarted.

It should be noted that if you were in a group chat with your friends who also use iPhone, either of the parties will need to start a new group chat to keep getting text messages. In a situation where you already removed your SIM card from your iPhone or you no longer have the old iPhone, you will need to get in touch with Apple to deregister the phone number associated with the iMessage.

You can simply get your sim deregistered using the screenshot shown below:

Transfer your SIM card to your iPhone.

Go to Settings.

Tap Messages.

Turn iMessage off. Go back to Settings.

Tap FaceTime.

Turn FaceTime off. Exit Settings.

iPhone's HEIF Photos (iOS 11 & Up)

You will need to save your HEIF photos to Google Photos from your iPhone in order to automatically see your original HEIF photos on your Pixel. To do this, you will need to backup photos and videos and also get photos saved which are taken using the device's camera as well as get photos saved from your phone to your Google Photos library.

1. Backup service: Using this service ensures that your photos and videos are saved to your Google photos library.

2. Privately stored: This means all photos and videos which are backed up from your device are private unless you decide to share them.

3. Syncing: By syncing, this will ensure that all changes are mirrored on every synced device.

Turning back up and sync on or off

1. Launch the Google Photos app

2. Get signed in to your Google Account.

3. Tap on Menu, at the top

4. Tap on Settings.

5. Select Back up & Sync.

If you are prompted with messages that ask you to allow the app to access your photos:

1. Launch the iOS settings

2. Select Privacy.

3. Tap on Photos.

4. Turn Google Photos on.

Changing Your Backup Settings

To back up your photos and videos, you will need to choose the Google account they are located by tapping the account name, under the Backup account.

To change your Upload size: Tap on Upload size and here you can change the quality of your videos and photos such that they fit your preferences or to get storage spaces freed up.

Storage for Pixel Users

If Google Photos backup have been set up on your Pixel phone, all photos and videos taken on your Pixel phone will be automatically backed up from your phone unto your Google photos.

Express Backup

If this option is not seen on your phone that means it is not available for you yet. Here are the features that come with the Express backup:

- They have faster backup over Wi-Fi as well as mobile data.

- It has unlimited storage such that Photos can be compressed to make use of less mobile data. With a photo that is larger than 3MP, such will be resized to 3MP.

- Good quality photos of 3MP can be printed to 6 inches by 8 inches size.

- All videos that are higher than 480p will get resized to a standard definition of 480p. Videos with 480p or less will look close to that of the original.

- Some information will be lost such as the closed captions.

IPhone And iPad Users

You will need to get the Google photos app downloaded and installed before getting started:

To get your upload sizes changed:

1. Open the Google Photos apps on your iPhone or iPad.

2. Sign in into your Google account.

3. Tap Menu at the top.

4. Tap on Settings.

5. Tap on "Backup & sync."

6. Tap on "Upload Size."

7. Tap on "High quality" or "Original quality."

Reducing the Size of Photos and Videos

In order to help save some space, you can change your already back-up photos and videos from the original quality to high quality. In a situation where you don't have any photos in original quality, this setting will not come up.

1. Go to photos.google.com/settings, on the computer.

2. Tap on Recover storage. It should be noted that this will not affect the quality of your future uploads.

3. Also, note that storage can only be recovered once a day. Also, photos or videos will not be changed to high quality if you have uploaded them with any of the following:

- Your Pixel
- Google Maps
- Google+
- Google Photos
- Google Drive
- Your Pixel 4 during the Original quality offer period
- Blogger
- Panorama

- Picasa Web Albums

- YouTube Web Channel Art

- Hangouts

Original Quality

All photos and videos get stored in the same resolution they were taken. For Google one users and the Google photos backup is set up, it should be noted that all photos and videos that are taken on your Pixel will be automatically backed up from your phone to Google photos.

Backup on Mobile Data

In a situation where you want to avoid waiting for Wi-Fi to get your Photos backed up, go to where you have "When to back up." Select the "Use cellular data to back up photos" or "Use cellular data to back up videos." It is to be noted that, depending on your service provider, charges from your service provider may apply if you are making use of mobile data in uploading.

Checking of your photos are backed up

1. Get the Google Photos app opened.

2. Sign in to your Google account.

3. Tap on Photos

4. Scroll to the top, you should see if your photos are backed up or is still waiting to back up.

Backing up Costs

You will have access to an unlimited free storage if your upload size is set to High quality but yours backed up storage will be counted against your Google Photos storage if it is set to original which will allow you to get up to 15GB of free storage of Photos as well as videos which are stored exactly as they are captured.

Other problems:

Don't have a cable

1. Tap Start on your Pixel phone, in a situation where you don't see the start, return to set up.

2. Tap Next

3. Tap No cable.

4. Tap "OK", under "Copy another way," then a backup from an Android phone.

5. Get both of your phones connected a Wi-Fi network.

6. On your Android phone, get the Bluetooth turned on.

7. Open the Google app to access the Google search.

8. Say "Ok Google, set up my device" or input in "Set up my device."

9. As soon as the name of your new Pixel phone shows up, tap on it.

10. You should then get a "Verify code" notifications on both phones.

11. Follow the onscreen instructions afterward.

Unable To Use the Old Phone

1. Ensure that there is a backup of your current android phone.

2. Tap Start on your Pixel phone.

3. Tap "Next," they can't use the old phone.

4. Tap Ok which you will find under "Copy another way," then a backup from the cloud.

5. Connect to Wi-Fi or Mobile network.

6. Select a Google account that has your data backed up in it and sign into it.

7. Follow the onscreen steps afterward.

Transferring without returning to set up or get selected data transferred only

Chapter 4: Copying Contacts

In a case where your Google account such as your Gmail has contacts attached to them, as soon as you sign in to your Pixel phone, these contacts will automatically be shown on the Pixel phone. To get contacts imported to your SIM card, you will need to get your names, email address, phone numbers, etc. saved to your Google contacts. Once contacts are saved in your Google account, they will get automatically synced with the Google contacts as well as all associated Android devices.

Using a Computer to Add Contact

1. Go to the Google contacts on your Computer.

2. Tap on Create account at the top left corner.

3. Input in the Contact's information.

4. Tap on "Save."

Importing contacts from a file

1. Go to Google contacts on your computer.

2. Tap "More," on the left

3. Tap "Import."

4. Choose how you want to import your contacts from the windows that come up.

5. Ensure your contacts are saved using extensions .CSV or .VCF file.

6. Click CSV OR vCard file

7. Select file.

8. Click on "Import."

From Gmail account

STEP 1: Exporting Existing Gmail Contacts

1. Go to Google contacts on your computer.

2. Click "More," on the left.

3. Click on Export.

4. Select the contacts you want to export.

5. Select Google CSV.

6. Click on "Export."

7. Click on your Profile picture and sign out at the top right.

STEP 2: Importing the file

1. Go to Google contacts on your computer.

2. Sign in using your Gmail account.

3. Click on "More" on the left.

4. Tap on Import.

5. Click on the "CSV" or vCard file

6. Choose your file.

7. Click "Import."

ISSUE: Can't import contacts

This problem does occur due to the following reasons:

• When more than 3,000 contacts are imported at a time. You are advised to split them into multiple CSVs before they are being imported.

• Another reason is when the number of contacts is over the 25,000 limits.

- Lastly, if the contacts are not formatted to work with the Google contacts. To avoid this, ensure that the contacts are saved as vCard or CSV file.

Controlling Who Gets Automatically Saved To Your Contacts

It is important to note that if you try emailing someone who hasn't been added as a contact, Google contacts will automatically save such person's email address into your "Other Contacts" group and their email address will automatically show up the next time you try emailing them.

Stop automatically saving contacts you email

1. Open Gmail on your computer.

2. Click Settings in the top right.

3. Select "I'll add contacts myself," in the "Create contacts for auto-complete" section.

4. Scroll to the bottom to click on "Save changes."

Seeing "Other Contacts" you have emailed

1. Go to Google Contacts on your computer.

2. Click "More" on the left

33

3. Click on "Other Contacts." It is to be noted that if you an iPhone, iPad or an Android user, the Other contacts option won't show up.

Deleting "Other Contacts" you have emailed

1. Go to Google accounts on your Computer.

2. Click on "More," on the left

3. Click on "Other Contacts."

4. Check the boxes that appear next to the names to select contacts.

5. Click on More and delete on the top right.

Adding Contacts on Android

1. Open the Contacts app on the Android phone or tablet.

2. Tap on the Add icon located at the bottom right.

3. Input in the contact's name as well as an email or phone number.

4. To pick which account in which you want to get your contact saved in: Tap the down arrow next to your email account.

5. To get more name details added: Tap on the down arrow next to the Name.

6. To get a photo added: Tap on add contact photo at the top.

7. Tap more fields to get more information such as street address or notes added.

8. Tap Save when you are done.

Importing Contacts

All your contacts can be added to your Google accounts after importing.

Your contacts will still be found in your other accounts.

From a SIM card

You can get all contacts saved on your SIM imported to your Google account by taking the following steps:

1. Get the SIM card inserted into your device.

2. Open the Contacts app on your Android phone or tablet.

3. Tap Menu at the top left.

4. Tap Settings.

5. Tap Import.

6. Tap the SIM card. In a situation whereby you have multiple accounts on your device, you will need to choose the account you like to get all the contacts saved in.

From A VCF File

All contacts saved to a VCF file can also be imported to your Google account.

1. Open the Contact app on your Android phone or tablet.

2. Tap on Menu at the top left.

3. Tap on Settings.

4. Tap Import.

5. Tap a VCF file. In a situation whereby you have multiple accounts on your device, you will need to select the account where you want all your contact saved in.

6. Search and select the VCF file to be imported.

Moving a Contact

Contacts can be moved from other accounts into Google accounts.

It should be noted that by moving an account will result in its deletion from the original account.

1. Open the Contacts app, on your Android phone or tablets

2. Tap on Contacts.

3. Tap Menu, at the top right.

4. Tap more and then move to another account.

5. Choose your preferred Google account you want the contact to be moved into.

6. Open your Pixel phone's Contacts app to see your contacts.

Chapter 5: Copying Music

Get your music backed up from your current phone into your computer.

With the aid of your Chrome browser, you can get songs dragged to your Google Play library.

Adding Music Using Google Play Music for Chrome

Music can be added to your Google play library with the use of Google Chrome by the use of the Google Play Music for Chrome.

You can easily get the Google Chrome downloaded on your Mac, Windows or Linux computer.

STEP 1: Check of your Chrome is updated

1. Open Chrome.

2. Select the More icon from the top right corner.

3. Then tap on Help.

4. Tap About Google Chrome

5. In a situation where an update is available, tap on "Relaunch." This will be found under the version number, once clicked updates will be installed.

STEP 2: Downloading Google play music for Chrome

1. Go to Google play music for Chrome.

2. Tap on Add to Chrome.

3. Tap on Add app.

STEP 3: Adding music to your Library

1. Open Chrome.

2. Ensure you are signed in to Chrome with the use of the same Google account used for your Google play music.

3. Go to the Google Play Music web player.

4. Tap on Google Play Music Menu icon.

5. Tap on Upload music.

6. You can choose to drag and drop files or select files to upload by choosing from your computer.

7. Get an update on the upload progress near the bottom of the screen

Note that if you are making use of iTunes can simply just drag and get music dropped from the iTunes library to the upload area.

Changing the Upload Settings

You can decide to change your music folders and also get your music updated. In order to do this, take the following steps:

1. Open Chrome.

2. Go to the Google Play Music Settings.

3. Click on Add your music.

4. You can then choose to either Review, add or remove folders that upload music to your Google Play Music Library.

5. Check the box next to "Keep Google Play up to date when I add new music to these locations," to get your library automatically updated when you add a song.

Having Trouble Getting Music Added: Troubleshoot Connection Issues

Playing music won't be able to establish a secure connection to Google if there should be an error with the connectivity.

The cause of the error in the connection may be due to the connection being blocked by anti-virus software, network administrator or some other network software.

To resolve the issue:

• Make use of a different Wi-Fi network in connecting your Play music if possible.

• Get the anti-virus software deactivated temporarily but make sure you remember to get it turned back on when you are done.

• Try contacting your network administrators as maybe they responsible for blocking the connection to Google.

Chapter 6: Transferring Music from a Computer to Your Phone or Tablet

You can access the Music library of your computer from your mobile phone or tablet with the use of one of the methods highlighted below after which you will then be able to use the Google Play Music app using any device.

Transferring Music to Your Online Library

To make your music available on any device you use with the Google Play Music app, you can simply upload music from your computer to your Google Play Music library.

Here are the following ways through which you can get music transferred from your computer to your Google Play Music library:

- Adding music with the use of Google Play Music for Chrome.

- Making use of the Music Manager software on the computer by getting it downloaded.

- Making use of the USB cable to load music unto your device.

- Loading music files from the computer to the mobile device using USB cable if it can be transferred online.

On a MAC

1. Get the Android file transfer downloaded and installed to your computer.

2. Get the screen unlocked if your screen is locked.

3. With the use of the USB cable, get your computer connected to your device.

4. Choose the Media Device (MTP), if there is a need to select a USB connection option on your device.

5. Go to the music files on your computer and get them dragged to your device's Music folder in Android file transfer.

On the Windows

1. Unlock your screen if it is locked.

2. With the aid of the USB cable, get your computer connected to your device.

3. Select the Media device if prompted to select a USB connection option on your device.

It is to be noted that music files will be needed to be downloaded if you don't have library already downloaded to your computer before getting music transferred onto your phone.

Using a Google Play Music Subscription

By subscribing to the Google Play Music, you will have access to stream or even download millions of songs without having to transfer the music library onto your device.

From your iPhone

In a case whereby you don't have a cable or want to make a transfer without going back to the setup, you can just go ahead and back up your iPhone as well as the iCloud account to your Google drive.

This will, in turn, enable your calendar, photos, contacts, and videos to be copied. Afterward, you will automatically find all your data on your Pixel phone.

The Google drive can be used for backing up the content on your iPhone and iCloud accounts.

The following will be backed up:

• Photos will be backed up to your Google Photos.

• Contacts will be backed up to your Google contacts.

• The calendar will be backed up to your Google calendar.

Things to Note Before Backing Up

Take note of the following before back up:

To get photos and videos backed up, your phone must have a working Wi-Fi network.

Only new photos will be back up if you do so multiple times, while also overwriting any previous backups for your contacts or calendars.

Even if you get your photos organized into albums, it should be noted that those albums won't get backed up to Google photos.

The contacts, as well as calendars from Facebook or Exchange services, will not get backed up.

Starting Your Back Up

1. Get the Google drive installed and opened on your iPhone.

2. Tap the Menu at the top left.

3. Tap "Settings" at the top.

4. Tap "Backup."

5. Tap on "Start Backup."

After You Are Done Backing Up

- All your content can be used across all devices as soon as you have backed up your contents.
- With the aid of the Google Photos, you can get to view and get your photos edited.
- Your photos will also get backed up to the original quality and in a situation where you run out of storage space, the remaining photos will get backed up in high quality.
- After turning on your Google contact sync on your iPhone or iPad, you will be able to see all your contacts.
- All your contacts from your iPhone and iCloud will now be seen in a new group which will be named after your device.
- Your calendar will now be viewable on your Google Calendar app.
- All your events will get added to a new calendar which will be named after your device.

What to do if the backup is incomplete

When all or some of your content are not completely backed up, you will be notified with a message that reads "Backup did not complete." If this happens to try the following:

• To avoid this, ensure that you update to the most recent version of the Google drive on your iPhone or iPad.

- Try backing up again as this may be a temporary issue.

- In a situation whereby nothing gets backed up, then you should check with the internet.

Transferring Selected Data Only

In your Google account: You will find your contacts in your communication apps on your Pixel phone, if you have already used them in your Google account, such as your Gmail.

From your iCloud account:
- Ensure all your contacts are in iCloud.

- All contacts should be exported from the icloud.com as a .vCard file.

- Get your contacts imported to Gmail.

Once your contacts have been copied, they will be found in the communications apps on your Pixel phone.

From your SIM card: In a situation whereby you have your contacts located on your SIM card,

- Get your Photos and videos copied.

• You can simply get your photos and videos backed up from your iPhone to Google photos. The Google photos will automatically appear on your Pixel phone.

- To view your photos and videos, get the Pixel phone's photos app opened.

Copying Music

Get your music backed up from your iPhone to your computer.

By dragging, move songs from your Chrome browser to your Google play library.

The music bought on from Google play will be automatically displayed on your Pixel phone.

Have your phone automatically recognize songs

This feature was originally introduced to the Pixel 2. It lets you access songs that are nearby and place their details on the lock screen.

1. Go to Settings.

2. Select Sound.

3. Tap Now Playing to turn it on.

View your Now Playing history and putting a shortcut on your home screen

1. Go to the sounds.

2. Tap Now Playing.

3. Tap Now Playing History.

You will be provided with a list of all the songs you have heard on your phone and the time you heard them. Play a song in apps such as Spotify, Play Music, and YouTube to play it. Place a shortcut for your Now Playing on the home screen so that you can easily access the feature.

Chapter 7: How to use your Screen

In this section of the book, we will be highlighting the various tips and tricks to familiarize you with the pixel phone.

Selecting and Moving Of Items

Tapping to Select: Tap on anything to select or start an action with on your pixel phone.

Tip: You can simply touch and hold a text so as to have the available actions that can be performed on it displayed such as playing a song or making a restaurant reservation.

Tapping to type: Tap on anywhere you want to type in by tapping. Once you type you will have the keyboard being displayed to perform this action.

How to Touch and Hold

On the screen, you can touch and hold items until the items respond, you can then release your finger.

How to Drag

i. Touch and hold an item.

ii. Then. move your finger on the screen without having your finger lifted.

iii. After dragging to a position of your choice, you can then release your finger.

For instance, you can get to move your fingers around on your home screen by dragging them with the steps above.

How to Swipe or Slide

In order to swipe or slide, all you have to do is to move your finger quickly across the surface of the screen without pausing. An example of this is to display other home screens by swiping the current home screen to your left or right.

How to Use Gestures on Your Device

The gestures are used in interacting with your pixel phone or Nexus device. Nevertheless, you have the choice to enable or disable some gestures.

To turn the gestures on or off;

i.   Visit the Settings app of the device.

ii. Tap on the "System" option.

iii. Tap on "Gestures."

iv. Select and tap on the Gesture you like to change.

Downloading Apps

i.   Visit the Play Store.

ii. Select your preferred iPhone apps for your Google Pixel phone.

Finding Recent Apps.

i.   Go to the bottom of the screen.

ii. Swipe up to the middle of the screen afterward.

Finding all your apps

i.   Go to the bottom of the screen.

ii. The swipe up to the middle of the screen twice afterward.

Chapter 8: How to Check Your Notifications

Checking phone by double-tapping

i. With your phone locked.

ii. Tap on the screen twice to have your notifications displayed.

Checking phone by lifting

i. With your phone locked.

ii. Pick your phone up to help in checking your notifications.

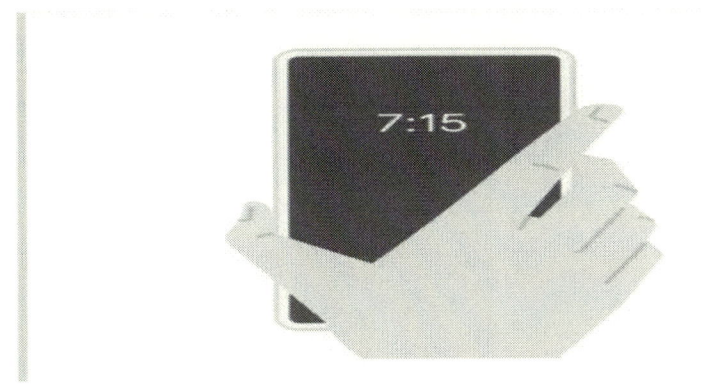

How to Control Notifications on Your Pixel Phone

The settings of certain apps or the whole pixel phone can be changed with a variation to the type of notifications you want. By swiping down from the top of your screen, you will have the notifications displayed and you can also get to see these notifications from the lock screen or home screen as well.

Note: Do Not Disturb can block notifications

The Do Not Disturb option can be used in limiting interruptions on your Android device.

The Do Not Disturb option also helps in silencing your Android device.

It can also be used in stopping vibration, muting sounds and also blocking visual disturbances.

You are allowed to choose what you like to block and allow for this action. You can make use of the Do Not Disturb option by taking the following steps though it should be noted that some of these steps will not work on the Android 9 or models that came after that.

i. In quick succession, have the interruptions turned off and on.

ii. Swipe down from the top of the screen to have the Do Not Disturb turned on or off.

iii. Tap on the Do Not Disturb

It should be noted that you can make use of the Smart Display or speaker with your Google Assistant in getting your phone silenced.

Flip To Shhh to Quickly To Enter Do Not Disturb

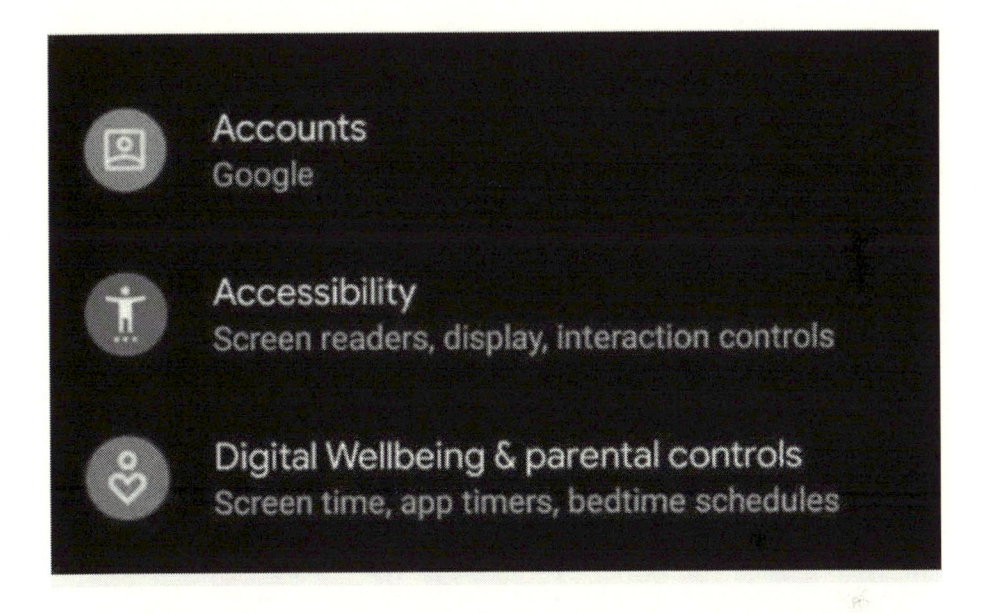

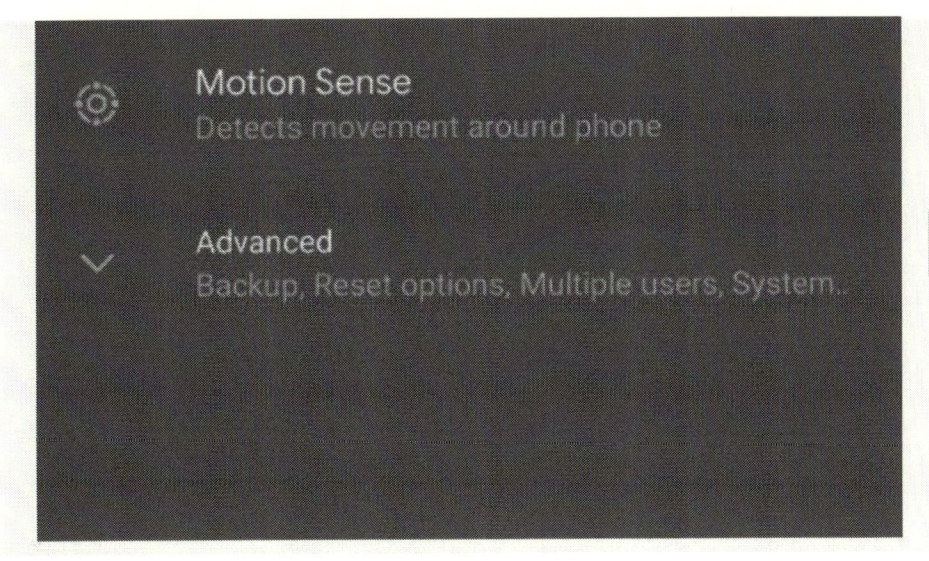

This feature helps to reduce distractions from your phone by switching your device to the Do Not Disturb mode when you place the phone on a flat surface face down. You can get this feature enabled by taking the following steps:

1. Go to Settings.

2. Tap System.

3. Tap Gestures.

4. Select Flip to Shhh.

5. Toggle on.

Show Notifications When You Reach For Your Phone

All you need to do here is wave your hand at your phone so you won't need the always-on display. The Pixel 4 is built with the Soli radar chip which detects your hand and turns your screen display on to give you details.

1. Go to Settings.

2. Select System.

3. Select Gestures.

4. Tap Motion Sense

You will be provided with the option to reach to check the phone.

Skip Tracks with a Wave of the Hand

You can skip tracks on your phone when playing music by waving your hand past the display of your phone. Simply wave your hand from the right to left to go forward and vice versa.

A glow will be displayed at the top to indicate the Motion Sense is detecting your hand.

You are provided with loads of options and controls by the Android Notification feature.

Direct Reply

You can now direct reply from any app as long as the feature is built into the apps.

1. Swipe down on any notification card.

2. Hit the "reply" option if is displayed.

3. You can now start typing without leaving that screen.

In some cases, you will be presented with the reply option by the toast notifications which enable you to reply while doing stuff like playing a game without taking your eyes away from the game.

Quickly switch to vibrate alerts

You can enable the vibration alerts by pushing the volume button and tapping the bell that pops up at the side.

Turn down media volume

1.	Hit the volume up or down button and the volume slider will be displayed at the right-hand side.

2.	Go to the bottom to tap the settings which will give you access to all the volume controls.

The media volume can be turned down from here.

Turn On Captions

This helps in adding captions to any speech in your video which comes in handy on occasions where you need your phone to be silent.

1.	Go to the side of your phone and hit the volume button.

2.	The caption toggle will appear at the bottom, tap on it to enable captions.

Squeeze To Silence Alarms and Calls

With just a squeeze, you can instantly silence your phone. You can enable this feature by taking the following steps:

1. Go to Settings.

2. Tap System.

3. Tap Gestures,

4. Select Active edge.

You will find the option to enable squeeze for silence at the bottom of the list.

Engage Do not Disturb

1. Swipe down to access Quick Settings.

2. Tap the Do Not Disturb icon.

This will spare you from all the pings and chirps every time you get likes from your Facebook.

Schedule Do not Disturb

1. Swipe down to access the Quick Settings.

2. Press and hold the Do Not Disturb button.

3. Select Schedule.

4. Turn on automatically and you will be provided with the automatic rules.

The times for Do Not Disturb can be set to automatically get turned on or off like in evenings or weekends.

To turn off notifications on an app

1. Go to Settings.

2. Tap Apps & notifications.

3. Select your preferred app.

You can block notifications for any apps by going to the notifications or when you are prompted with a notification that you don't want, swipe slowly to the right to access a settings cog. You can block notifications for the app here.

Hide sensitive information in lock screen notifications

1. Go to Settings.

2. Select Display.

3. Select Lock screen display.

4. Tap lock screen.

5. Select the "show sensitive content only when unlocked,"

This will make your phone hide information so that intruders will not have access to read them.

Chapter 9: Make and Receive Calls on Your Phone

You can use the following apps to make phone calls:

• Phone app

• Contacts app

Or any other widgets or apps that is displayed in your contact app. Tap a phone number to dial it. You can also copy underlined phone numbers from Google chrome to the dialpad by taping the number.

How to Make A Phone Call

1. Open the Phone app on your phone.

Pick the contact you want to dial:

2. You can also tap the dial pad keyboard to enter a phone number. You can also tap Contacts to choose a saved contact. The app may display list of your frequent contacts.

1. Tap Recents to choose a phone number you recently contacted.

2. Tap Speed dial to choose a phone number saved to your speed dial.

3. Then tap Call. To end the call, tap End call.

You have to drag the call bubble to the bottom right of your screen if you minimized the call. Some devices and carriers may also support RTT calls and video calls.

To pick a call, when your phone is locked, you'll have to swipe the white circle to the top of the phone screen or tap Answer.

To decline a call, when your phone is locked, you'll have to swipe the white circle to the bottom of the screen or tap Decline. To Dismiss the call and message the caller, tap the Message icon.

If voice mailbox is activated on your phone, you can decide to drop a message from rejected callers.

Use Phone Call Options

While on a call:

1. To locate the keypad, press the keypad icon.

2. Tap Speaker to switch between speakerphone, earpiece, or any connected Bluetooth device.

3. Tap Mute, to mute or un-mute your phone's microphone.

4. Tap the Hold button to temporarily suspend the call without ending the call.

5. Tap the Hold button again to return to the call.

6. Tap the Switch button, and the current calls will be put on Hold automatically.

7. Tap Add-call to join more contact into a conference call.

8. Tap the Home button to minimize the current call or drag the call bubble to the Home button.

9. Drag the call bubble to the down part of the screen to Hide.

10. Tap Video call/Facetime to switch to a video call once the phone runs on Android 6.0 or above.

Make Calls Over Wi-fi

Instead of making a call using your mobile network, you can as-well call through Wi-fi connection. The only downside is that not all phone carriers support calls over Wi-fi.

Note: Additional charges may be incurred for making calls over Wi-fi. Since not all phone carriers support call over Wi-fi, you will have to Open the settings on the device to check the phone service provider for more details about the phone.

How to turn on Wi-fi call

In order to make calls over Wi-fi, you will have to Open the WiFi Settings and turn on its features.

If Wi-fi call is not supported by your phone carrier, you can decide to make use of a voice over IP service.

Step 1

1. Open the Phone app.

2. Open Settings

3. Click on calls

4. Click on Wi-Fi. If these options cannot be found on your device, it simply means your phone has no access to Wi-Fi calls

Step 2: Use VOIP service to make calls via Wi-Fi (SIP)

Yes, Some devices can be used to make calls over Wi-Fi, such as a voice over IP service. In order to activate that on your device, you need to add a "Session Initiation Protocol" (or SIP) account on your device.

1. Click on phone Settings.

2. Tap more, in order to bring the remaining settings options.

3. Locate Calls and then Tap Calling accounts And click on SIP accounts.

4. Then Tap Create.

Fill the requested bio-data information like username, password, and server of your new SIP account.

Pro Tip: In order to modify your nickname, port number or some other required fields, click Optional settings, and then Tap Save.

Check Your Voicemail

You can make a call to your voicemail service. With some devices and carriers, a list of your voicemail notification can be seen on your phone. Once your device isn't connected to Wi-Fi, your calls will automatically use your mobile network, if any, is available.

Ways to check your Voicemail

1. Open message notification

2. Tap voicemail notification, then you can view and listen to your message on your device.

3. Then call your voicemail service to check the missed messages.

Voicemail Call

You can use Voicemail service for calls, and also to check on messages:

• Tap your phone app.

• Tap Dialed Keypad at the bottom of the screen.

- Touch and hold 1.

You can check how voicemail is listed on your device, via your phone app. But not all carriers can offer this feature.

- Open Settings and turn on visual voicemail.

- Open the phone app on your phone.

- Tap More at the right top part of the screen.

- Click on settings and then Voicemail.

If the visual voicemail is off, it will erase the recording from your phone app, but your phone carriers keep a copy of the recording.

1. See a list of your voicemails in the Phone app.

2. On most phones, voicemail is usually listed on the Phone app, but not all carriers obtain this feature.

3. Open the phone app.

4. Swipe down the screen to check voicemail notification. If "voicemail" cannot be found there, then call voicemail instead.

How to share the Voicemail Audio Recording?

All you need to do is to tap Send in order to send. Then click on the app that you need to use. Your voicemail transcript can be seen if Xfinity or T-Mobile is your service provider.

Transcription is only accessible on Android 8.0 and above in English and Spanish.

How to Check your Android Version

1. Open Settings and On voicemail transcription

2. Click on more in Settings and Tap Voicemail.

3. Locate and then ON the voicemail transcription.

Note: Be sure that the visual voicemail is turned on In case you don't see voicemail transcription.

Tip: To share a transcript to another person, click on share and then the app.

Note: Google will automatically transcribe old and new voicemails record when its turned on. Google uses computers to transcribe voicemails and doesn't associate them with your Google account.

If voicemail transcription is turned off, the recording list and transcription are deleted on your phone app.

You can denote Voicemail transcripts and recording lists in order to help improve transcription technology. The voicemails that are donated may be reviewed by humans but will not be associated with your Google Account or phone number.

How to turn on voicemail transcription donation:

• Open Settings on your phone.

• Click More settings

•Turn on Voicemail transcription analysis.

How to change your voicemail settings:

• Open Settings

• Click More, at the top right.

• Click on Settings and then Voicemail.

The settings you can do on Voicemail

1. You can Change what handles your voicemails: Tap Advanced and then Service.

2. You can Set up your voice mailbox:

Tap Advanced and then Setup.

3. You can change your notification settings: Tap Notifications.

4. You can Turn on vibration: Tap Notifications and then Vibrate

66

What Can Cause Voicemail Notification Not To Work

If you switched off a phone that lists voicemail on the phone app to one that doesn't, voicemail notification may not be gotten. If voicemail list doesn't come up in your phone app:

Step 1: Contact your mobile service provider. Request for a downgraded version from your mobile service provider to a basic voicemail service.

Step 2: Check your voicemail settings

1. Open your phone app.

2. Tap more, at the top right.

3. Click on settings and then tap Voicemail then Advanced.

4. Click on Service and be sure your device carrier is selected.

Change Call Settings

If you want to change your phone's call ringtone, call history display, vibration settings, and quick responses You can. All you have to do is to:

1. Open the phone app.

2. Click More And then Settings.

3. Tap Sounds and Vibration.

4. Tap phone ringtone, to select from available ringtones.

5. Tap vibrate for calls, to make your phone vibrate when your Phone's ring.

6. Tap keypad tones to let your phone sound anytime you press the keypad.

Change Caller Name Display

You can change Callers' names, and format, and list them the way you want on your phone. To do that:

• Open the phone app.

• Click More And then Settings.

• Click on Display options.

• Tap Sort by, to select how you want your phone to sort calls in your call history.

• Tap Name format to choose how your phone shows contact names in your call history.

Change Text Responses

You can send an automatic text message instead if you can't pick up a call. These procedures listed below will guide you to change your automatic text messages:

- Open the phone app.

- Click on Settings.

- Tap More.

- Click on quick responses.

- Tap a response from the list display.

- Edit the response.

- Tap Ok.

Use TTY or RTT with calls

Text support can be added to your phone calls.

Note: check with your carrier if you are not sure whether you can use TTY or RTT

1. Open the phone app.

2. Click on Settings.

3. Tap Settings And then More.

4. Click on Accessibility.

TTY mode will be displayed, tap it.

You can decide to Off TTY to hear and speak on the phone without text support.

1. Tap TTY Full, to use typed text in both directions.

2. Tap TTY HCO, to type text and hear the other person reply aloud.

3. Tap TTY VCO, to read and speak aloud the other person's reply as text.

4. Turn on the Switch if Real-time text shows up.

Turn Off Nearby Places Search

You can search for Nearby places via your phone app, even if they are not on your contact list. For you to do this, you will have to permit your phone to use your current location.

How to turn off nearby places:

1. Open the phone app on phone.

2. Tap Settings then More And then Nearby places.

3. Tap Nearby places at the top of the screen.

Note: if you see location permission is denied under "Google Account," Tap it.

How to set your phone to make calls with contact when you travel outside your home country.

You can make calls with your phone app when you travel outside your home country. Your phone app will automatically add the correct country code to a stored contact.

Your home country will be automatically set by your phone app. To change your home country:

1. Open the phone app.

2. Click on Settings.

3. Tap More and then Settings.

4. Tap Calls and then Assisted dialing.

5. Click on the Default home country.

6. Choose the country you currently are.

7. Tap Save.

To stop your phone from adding a country codes automatically;

1. Open your phone app.

2. Tap Settings.

3. Click on More And then calls.

4. Click on Assisted dialing.

5. Tap Default home country.

6. Turn off, Assisted dialing.

7. Use spam and caller ID protection.

When the spam protection is on, and you make or get a call with caller ID, you can see the callers information.

Your phone may need to send information about your calls to Google if you are to use caller ID and spam protection.

How to Turn Caller ID & Spam Protection Off Or On

Both are on by default, but you can decide to turn it off. Your phone will need to send information to Google about your calls for you to use caller ID & spam protection.

1. Open your phone app.

2. Click on Settings.

3. Tap More and then Caller ID & spam.

4. Turn on or off Caller ID & spam.

Optional: You can turn on Filter suspected spam calls to prevent spam calls from ringing on your phone. The only thing you will see on your notification list is filtered calls, and you will be able to check any received voicemail. But you won't get voicemail notification or missed calls.

How to block or report spam on your phone

1. Open your Phone app.

2. Tap Recent calls.

3. Click on the call you want to block or report as spam.

4. Tap Report spam / Block.

6. You will be asked if you really want to do any of the two options.

If you select Report as spam.

7. Tap okay.

Recognize Spam

You can block and report any number if the number looks suspected as spam caller or "spam" as the caller ID.

How to report a mistake if someone you know is marked as spam

1. Open your device's Phone app phone.

2. Click on Recent calls.

3. Locate the contact mistakenly added as spam.

4. Tap Unblock number

5. Tap Unblock.

How to Block or Unblock A Phone Number

You can block some numbers if you don't want to pick their calls. Your phone will automatically reject the calls when the numbers try to call you.

1. Open your Phone app.

2. Click on Call history.

3. Locate the number you want to block.

4. Tap Block /Report Spam.

Note: you can still get a voicemail from blocked callers if your phone doesn't have a visual voicemail.

How to Unblock a number

1. Open your Phone app.

2. Click More

3. Click on Settings and then Blocked numbers.

4. Blocked numbers will be displayed, and then tap Unblock.

You can see who is calling and why before answering a call. Your phone uses Google Assistant to screen calls, and that's what makes it easy to discover who is calling and why is the person calling before you can answer a call.

Note: For some Android phones and all pixel phones, you can screen calls in English, Canada and the United States.

1. Check the Phone app if it's Updated, if it's not, tap it to update the app.

2. Tap Screen call when someone calls.

3. Your call will be screened by Google Assistant and ask who's calling and why.

A real-time transcript of how the caller responds will be shown to you.

Here are some responses the caller will hear:

Is it urgent? - "Do you need to get a hold of them urgently?"

Report as spam - "Please remove this number from your mailing and contact list. Thanks and goodbye."

I'll call you back - "They can't talk right now, but they'll give you a call later. Thanks and goodbye."

I can't understand - "It's difficult to understand you at the moment. Could you repeat what you just said?"

How to Get Transcripts for Screened Calls

1. Locate your screened call,

2. Tap the contact's name or number.

3. And Click Call details. This will bring up a screened call.

4. Then tap Transcript.

Information to take note: Third-party app doesn't work with Call screen, call recording, and screen recording apps. We recommend to turn off these apps before Call screen can be used.

Text messages

How to Change Your Interruptions Settings

By activating the Do Not Disturb⊖ option, sounds, and vibration on your device will be automatically stopped. Interruptions such as the alarms, notifications, messages, and calls can be configured as regards the way they act.

How to Set What To Block

It should be noted that Settings do vary depending on the device, so you are required to contact the device manufacturer for more information.

i. Open the Settings app on your device.

ii. Tap on the "Sound" option.

iii. Tap on "Do Not Disturb."

iv. It should be noted that if you are making use of an older Android version, you will have the "Do Not Disturb preferences" come up instead.

v. Go to the "Behavior" option to choose what you want to allow or block.

vi. You will also have the Sound & Vibration option where you will be able to block or allow your media, alarms or touch sounds.

vii. The sounds from notifications, sounds or visuals can also be blocked or just pick the custom options. Also, when you pick up your Custom option, you will have the opportunity to customize what is displayed when your screen is on and what also happens when you have the screen off.

Critical notifications will be displayed using a setting. The System security notifications can't be blocked.

How to Set Exceptions to Allow

To set the exceptions, you are required to take the following steps:

i. Open the Settings app of your device.

ii. Tap on the Sound option.

iii. Tap on the Do Not Disturb option.

iv. In a situation where you are making use of an older Android version, you will have the Do Not Disturb preferences come up instead.

v. Choose what to allow under the "Exceptions" option.

vi. Allow calls by taking the following steps:

- Tap Allow calls.

- Pick whose calls to allow by choosing from the following options: anyone, contacts only or starred contacts only.

- To stop calls from coming through, tap on the "None" option.

- In a situation where you want to allow the call of the same person if such person calls twice within 15 minutes, turn on the "Allow repeat callers."

vii. You can allow messages by taking the following steps:

- Tap on the Allow messages option.

- Pick whose messages to allow by choosing from the following options: anyone, contacts only or starred contacts only.

- To stop messages from coming through, tap on the "None" option.

viii. Turn on the Allow reminders option to allow the reminders.

ix. To allow events, turn on the Allow events option.

How to Set How Long The "Do Not Disturb Option" Will Last

i. It should be noted that these settings do vary depending on your device.

ii. Open the Settings app of the device.

iii. Tap on the Sound option.

iv. Select the "Do Not Disturb" option.

v. Then you can select the Duration from here.

vi. In a situation where you are making use of an older Android version, you will have the Do Not Disturb preferences come up instead.

vii. Now you can pick the Duration of how long Do Not Disturb option will last except your device is generally turned off.

How to Stop Interruptions Automatically

To have the Do Not Disturb option automatically turned on at certain times, you have the options to set certain rules. To stop interruptions automatically, you can take the following steps:

i. Open the Settings app of your device.

ii. Tap on the Sound option.

iii. Tap on the Do Not Disturb option.

iv. Select to turn on automatically.

v. In a situation where you are making use of an older Android version, you will have the Do Not Disturb preferences come up instead.

vi. You can tap on a rule or select your own rule by selecting to add a rule and time.

vii. Then you can have your rule's name, status as well as the alarm override name edited.

viii. To have a rule deleted, you will choose the Delete option.

How to Set Your Do Not Option during Events & Meetings

You can set event rules on your Google calendar to automatically turn on the Do Not Disturb option during some of your events by taking the following steps:

i. Open the Settings app of the device.

ii. Tap on Sound.

iii. Tap on the "Do Not Disturb" option.

iv. Turn on automatically.

v. In a situation where you are making use of an older Android version, you will have the Do Not Disturb preferences come up instead.

vi. You can tap on an event or select your own rule by selecting to add a rule and then Event.

vii. You can have the rule edited such as the naming of your rule afterward.

viii. In a situation where you have multiple accounts with the Google Calendar events, you can pick one of this during events.

ix. Check from the top to be sure your rule has been turned on.

x. Tap on the Delete option to have a rule deleted.

Using An Android 8.1 And Below: Picking Which Interruptions To Allow

Option 1: Total Silence

To complete place your phone in the silence mode such that it doesn't make any sounds or even vibrate, you will have to select the "Total silence."

i. Go to the top of the screen and swipe down with the use of your two fingers.

ii. Go to the Do Not Disturb option or from your current option and select the Down arrow.

iii. Have Do Not Disturb option turned on.

iv. Select "Total Silence."

v. Select the duration of this setting.

vi. Tap on "Done."

vii. You will then see the icon to indicate total silence.

It should be noted that once the Total silence ⊖ is turned on, the following will happen to your device:

• The device will not vibrate or make any sounds when there is a call, notification or message.

- Sounds from your videos, games, music or other media will be turned off.

- There will be no sounds from your alarms.

- Nevertheless, whenever you have a call you will still be able to get receptions.

Option 2: Alarms Only

By picking the Alarms only, you will still be able to receive sounds from your videos, games, music or other media with your phone will still be placed in the mute mode.

To select the alarms only, you are required to take the following steps:

i. Go to the top of the screen and swipe down using your two fingers.

ii. Go to the Do Not Disturb option or from your current option and select the Down arrow.

iii. Turn on the "Do Not Disturb" option.

iv. Select the Alarms only.

v. Select the Duration of your preferred settings.

vi. Tap Done.

vii.   The Alarms only Do Not Disturb⊖ will be displayed

Option 3: Priority Notifications only

The Priority Only option will only allow you to be interrupted by important notifications. The sounds from your videos, games, music or other media will remain intact and won't be muted.

To pick the Priority only option, you are required to take the following steps:

i.   Open the Settings app of the device.

ii.   Tap on the "Sound" option.

iii.   Tap on the Do Not Disturb preferences.

iv.   Select the Priority only to allow.

v.   Set your priorities.

vi.   Go to the top of the screen and swipe down using your two fingers.

vii.   Go to the Do Not Disturb option or from your current option and select the Down arrow.

viii.   Turn on the "Do Not Disturb" option.

ix.   Select the Priority Only.

x. Choose the duration of your settings.

xi. Tap Done.

xii. The Priority only Do Not Disturb⊖ will be displayed.

How to Automatically Block Interruptions

Option 1: Silencing the sounds at certain times

If you like automatically have your phone in a silent mode, such as during bedtime, you can simply set some rules to have this regulated:

i. Open the Settings app on your device.

ii. Tap the "Sound" option.

iii. Tap the Do Not Disturb preferences.

iv. Go to the "Automatic rules."

v. Tap on a rule such as Weeknight,

vi. You can choose to make your own rule by tapping Add more and the Time rule.

vii. You can choose to Edit your rule.

viii. Check the top of the screen to ensure your rule is turned on.

Option 2: Silencing sounds during events and meetings

You can set event rules to help automatically silence your device during occasions like meetings or events. To do this, you are required to take the following steps:

i. Open the Settings app of the device.

ii. Select Sound.

iii. Select the Do Not preferences.

iv. Go to the Automatic rules.

v. Tap on Event to have the default rule edited.

vi. Tap Add more and then click on the Event rule to make your own rule.

vii. Edit your rule.

viii. Check to ensure your rule is turned on by scrolling to the top of the screen.

Option 3: How to Block Visual Disturbances

To avoid being visually interrupted by silenced notifications, you are required to take the following steps:

i. Open the Settings app from your device.

ii. Tap on the "Sound" option.

iii. Tap on the Do Not Disturb preferences option.

iv. Tap on "Block visual disturbances."

v. Turn on Block when the screen is on so as to prevent notifications from showing over your screen when doing something else.

vi. Turn on the Block option when the screen is off so as to prevent notifications from pulsing light or turning on, on the screen.

How to Turn Interruptions Back On

A. Turn off Do Not Disturb

To have the Do Not Disturb option turned off, you can do one of the following:

Go to the top of the device screen and swipe down, after which you can choose from the following options: Alarms only⊖, Total silence⊖ or Priority only⊖

OR

Press the Volume down button and tap on the Turn off now.

B. Overriding Do Not Disturb for selected apps

To enable notifications for certain apps, you are required to take the following steps:

i. Open the Settings app of the device.

ii. Tap on the "Apps & notifications."

iii. Select "App."

iv. If the app does not come up, tap "See all apps" or "App info," and select the App option from there.

v. Select "App notifications."

vi. Turn on the Override Do Not Disturb.

vii. In a situation where you don't see "Override Do Not Disturb," Tap on the Additional Settings in the app.

viii. Tap on Notifications.

ix. You can then turn on Override Do Not Disturb from here.

How to Control notifications on your pixel phone

To snooze a notification, take the following steps:

i. Drag the notification slightly to the right or left.

ii. Select the Snooze.

iii. Select the Down arrow to pick a time.

To clear one notification: just simply swipe in the right or left direction.

To have all the notification cleared:

i. Scroll to the bottom of your notifications.

ii. Select "Clear all."

It should be noted that to clear some notifications such as the music player notification, the music has to stop playing first.

How to Reply, Archive, Expand And More On Your Notification

• Select an action like the Reply or Archive to perform an action directly on a notification.

• Select the Down arrow to have your notification expanded.

• Select the app that sent the location, while on the lock screen, you will have to double tap.

It is to be noted that you can only tap on your oldest notifications from the notification dots and clear the notifications one after the other to have the next one displayed.

How to Control your Emergency Broadcasts

You can edit your emergency alert settings such as the AMBER alerts and threats, by taking the following steps:

i. Open the Settings app from your device.

ii. Select "Apps & Notifications."

iii. Select "Advanced."

iv. Select "Emergency alerts."

Chapter 10: Messages

You can receive and send messages, voice messages, photos & animation, and video using Messages.

How to Make Messages Your Default App

You can make Messages your default if your phone has multiple messaging apps.

1. Open the Messages app.

2. Follow the onscreen instructions.

Messaging app can also be set as default from your device's Settings app.

Start Sending and Receiving Messages with Your Contact

If you store any number on your phone, it will show as a contact in your Message app. And another new contact can be added within the app itself.

1. Open the Message app.

2. Tap start chat.

3. Enter the receiver phone number, name, or email.

4. Type your message.

5. After you have type what you want then

6. Tap Send.

Add a new contact on your conversation list.

1. Open the Messages app.

2. Find and select the number you wish to add to your contact on your conversation list.

3. Click on More icon.

4. Then, tap Add contact.

Add A New Number as A Contact from A Group Conversation

1. Open the Messages app.

2. Open the group conversation and select the number you want to add.

3. Click on the number you want to add and then Add contact.

Change notification settings for specific people

1. Open the Messages app.

2. Open a conversation with the person you want to change a notification when they message you.

3. Tap More Option and then Notifications.

4. Notification can be turned on or off completely.

Block a Contact

When a number is blocked on your device, the number will not have access to call or text you a message.

To do that:

1. Open the Messages app.

2. Click on the contact you wish to block

3. Tap More Options.

4. Tap Block.

You can receive and send videos, audio files, stickers, GIFs, as multimedia messages (MMS) in messages.

To send videos, Files or GIFs,

1. Open to the Android Messages app.

2. Start or open a conversation.

3. Tap Attach.

Choose or select whatever media you want from your phone's file & videos, click photos, GIFs or document files.

You can also take a fresh picture from your phone camera or start voice recording.

1. Click on the file you want to send on the list.

2. Then tap Send.

Note: Recordings or photos that you take in Android Messages will not be saved on your device.

Send voice messages

1. Open Android Messages app.

2. Start or open a conversation.

3. Press and hold the Microphone sign at the right-hand side of the text box.

4. Record your message.

5. Release the Mic to Send.

7. To cancel your voice message, hold the message and an option will show up on the screen to delete.

Send Your Location

1. Open the Android Messages app.

2. Start or open a conversation.

3. Tap Attach at the left-hand side of the text box.

4. Select and tap Location.

5. Tap send your Location.

6. Tap Send.

Change Your Messages Notifications and Settings

Your google Pixel 4 gives you the option to customize your notifications to suit your preference. You can customize your device to play sounds, vibrates or sends reminders when a message comes in.

Change Global Settings

This feature gives you the option to change options for specific people.

1. Open the Messages app.

2. Tap More and then Settings.

3. Change the phone default for messaging app: Click on the Default SMS app.

How to change what happens on your phone when you get a message:

1. Tap Notifications and then Importance.

2. Turn off sounds playing whenever you send messages: Turn off Hear outgoing message sounds.

If you want to be notified on your phone whenever you used Messages for web,

1. Open your settings app.

2. Tap Apps & notifications and then Messages and then App notifications.

3. Turn on Get notifications while using the web.

Change Advanced Settings

You can change how your Messages app sends text, pictures, videos, & voice recording.

1. Open the Android Messages app.

2. Tap Settings, and then More.

3. Tap Advanced.

Send a message or files separately to each person in a conversation: Tap Group messaging and then Send an SMS reply to all recipients and get individual replies (mass text).

Files can be downloaded in Messages when you get them: Turn on Auto-download MMS. Tap Auto-download MMS while the data started roaming.

You can change your settings on how you want reports to be delivered, or how you want to send phone number.

1. Open the Messages app.

2. Tap More and then Settings and then Advanced.

3. To change Simple, special characters in text messages: Turn on Use simple characters.

If you would like to get delivery report anytime you send a message;

1. Turn on Get SMS delivery reports. (Delivery reports depend on your carrier.)

2. Have the number you use to send files changed:

3. Then Tap Phone number.

Change options for specific people

1. Open the Messages app.

2. Then have a conversation with one of your contacts opened.

3. Tap More,

4. Then tap People & options.

How to Stop Getting Message Notifications outside Messages for a particular person: you will have to Tap Notifications and then Turn off Notifications.

How to change what happens on your phone when you get a message from a person: You will need to Tap Notifications and then Importance.

How to Stop a number from sending you a message: You will have to Tap Block [phone number].

Adding a new member to a group conversation: Tap Add people. Exit a group conversation: You can access this feature if you turn on the chat feature— Tap Leave group.

Turn On Dark Theme

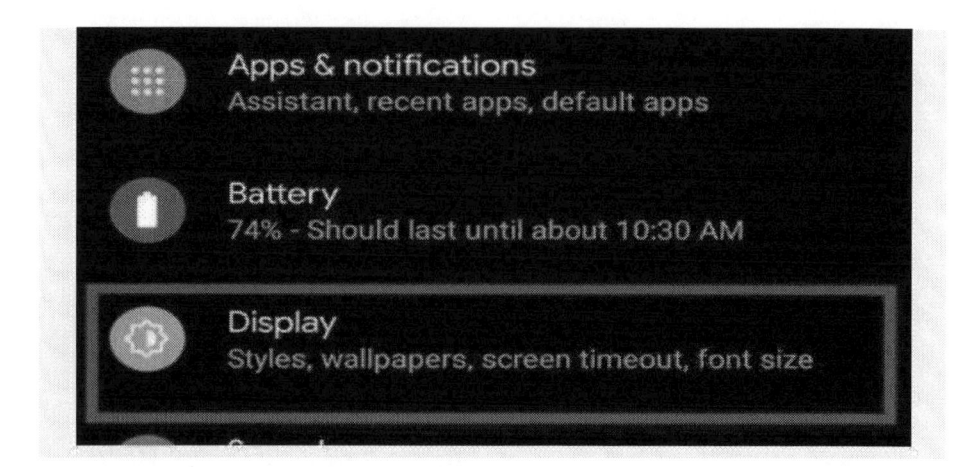

Apps & notifications
Assistant, recent apps, default apps

Battery
74% - Should last until about 10:30 AM

Display
Styles, wallpapers, screen timeout, font size

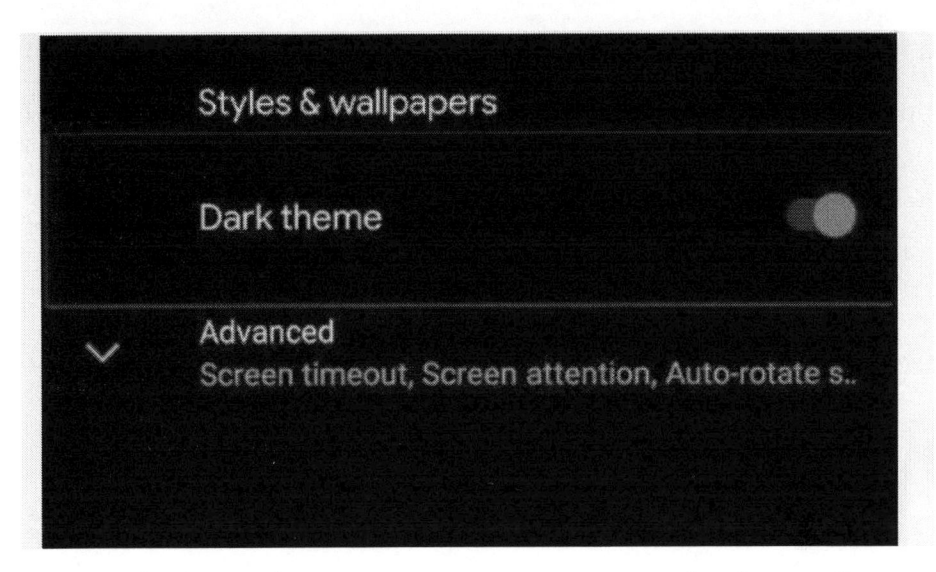

> Styles & wallpapers
>
> Dark theme
>
> ⌄ Advanced
> Screen timeout, Screen attention, Auto-rotate s..

The Dark theme feature can serve as a great visual complement especially at night, making your device screen easy on the eyes irrespective of the color of your Pixel 4. This feature has been around for a while and it has also been introduced into Android 10.

Using this feature is a great way of saving your battery life.

Note: This step can only run on Android 10 and above.

On your mobile phone:

1. Open your phone's Settings app.

2. And then, Tap Display and select Dark theme.

3. Tap Okay.

4. Open Quick Settings by swiping down from the top of your phone screen and then Turn on Dark theme.

5. The setting must firstly be added to Quick Settings.

Turn off Verified SMS

Google verifies messages from participating businesses when Verified SMS is turned on.

The steps below will guide you on how to put it off:

1. Open the Android Messages app.

2. Click on More.

3. Then tap Settings.

4. Tap Verified SMS.

5. Turn the switch to the left "Verify business message sender."

Note: At this time, Verified SMS only works when chat features are enabled and it is only available in select countries: United States, France, Canada, Mexico, United Kingdom, India, Philippines, Spain, and Brazil.

Chapter 11: Choose How You Get Notifications On Your Device.

i. Open the Settings app from your device.

ii. Select "Apps & Notifications."

iii. Select Notifications.

iv. Select the options you like to use as the defaults of your device.

v. On the lock screen, you can:

- Allow the notification dots Blink light

- Default notification sound

- Do Not Disturb

- Also, you can have your screen woken up when a notification comes in.

How to change the notifications for certain apps

Option 1: From the Settings app

i. Open the Settings app of the device.

ii. Select Apps & notifications.

iii. Select Notifications

iv. Go to the "Recently sent" option to view the apps that just got notifications sent to you.

v. You can perform the following actions:

• Turn off all notifications available for the listed apps.

• Select the app's name to assign specific categories of notifications.

• Tap on the "See all from last 7 days" to see more apps. Tap on these apps to help sort by "Most recent or most frequent."

Option 2: On a notification

i. Swipe down from the top of the screen.

ii. Drag the selected notification to the right or left direction slightly.

iii. Click on the "Settings ⚙" option.

Or you can simply touch and hold the app from the notification dot and tap on the info ⓘ icon.

Option 3: From the app itself

By visiting the settings menu of the app itself, you can get to control its notification like having control on the sound notification from the app's settings.

How to Turn Notification Dots on Or Off

In some cases, apps do show a dot when you get notifications from them can touch and hold these apps using the dot to see the notifications. To have these notifications turn on or off, you are required to take the following steps:

i. Open the Settings app ⚙ from the device.

ii. Select Apps & notifications.

iii. Tap on Notifications.

iv. Then turn the Allow notification options on or off.

How to Control How the Notifications Display On Your Lock Screen

Option 1: Don't show the notifications

From the lock screen, you can get to block all notifications, which will result in you seeing your notifications only when your device is unlocked.

i. Open the Settings app ⚙ of your device.

ii. Select Apps & notifications.

iii. Tap on the "Notifications."

iv. Tap on the Lock screen

v. Select the "Don't show notifications at all."

It is to be noted that once you make this option your default, it will automatically apply to all apps.

Option 2: Showing complete notifications

By default, you will be able to see all the contents of your notification from the lock screen.

i. Open the Settings app ⚙ from the device.

ii. Select Apps & notifications.

iii. Select the "Notifications" options.

iv. Tap on the lock screen.

v. Tap on the "Show all notification content" option.

It is important to note that some apps on your lock screen can be restricted even if you make this option as your default.

You can set some apps to hide sensitive contents of your notification or shouldn't show notifications at all on the lock screen.

Option 3: Show notifications but hide sensitive contents

i. Open the Settings app of your device.

ii. Select Apps & notifications.

iii. Select Notifications

iv. Tap on the lock screen.

v. Tap on "Hide sensitive contents.

It is to be noted that once you turn on lockdown, the notifications won't show up on your lock screen.

Chapter 12: How to Set Your Pixel Phone to Automatically Unlock

In some cases like having your phone in your pocket or near home, you can choose to keep your phone unlocked.

Features that can be used basically depend on your device, like when making use of the Smart lock, there won't be the need to unlock with PIN, password or pattern.

Turning the automatic unlock on will change your screen lock.

i. Open the Settings app of the device.

ii. Select the Security & location.

iii. Select "Smart Lock."

iv. Input your PIN, password or pattern.

v. Select an option.

vi. Follow the on-screen instructions.

It is to be noted that when you have the screen turned on after setup, a pulsing circle will be displayed at the bottom close to the lock icon.

Also in a case where your device has been idle for 4 hours after it was restarted, you will need to get it unlocked.

How to turn off automatic unlock

i. Open the Settings app ⚙ of the device.

ii. Select the Security & location.

iii. Select the Smart lock.

iv. Input your PIN, password or pattern.

v. Then have the On-body detection turned off.

vi. Remove all trusted devices, trusted faces, trusted places and Voice match.

vii. You can choose to turn off your screen lock.

Skip the lock screen with face unlock

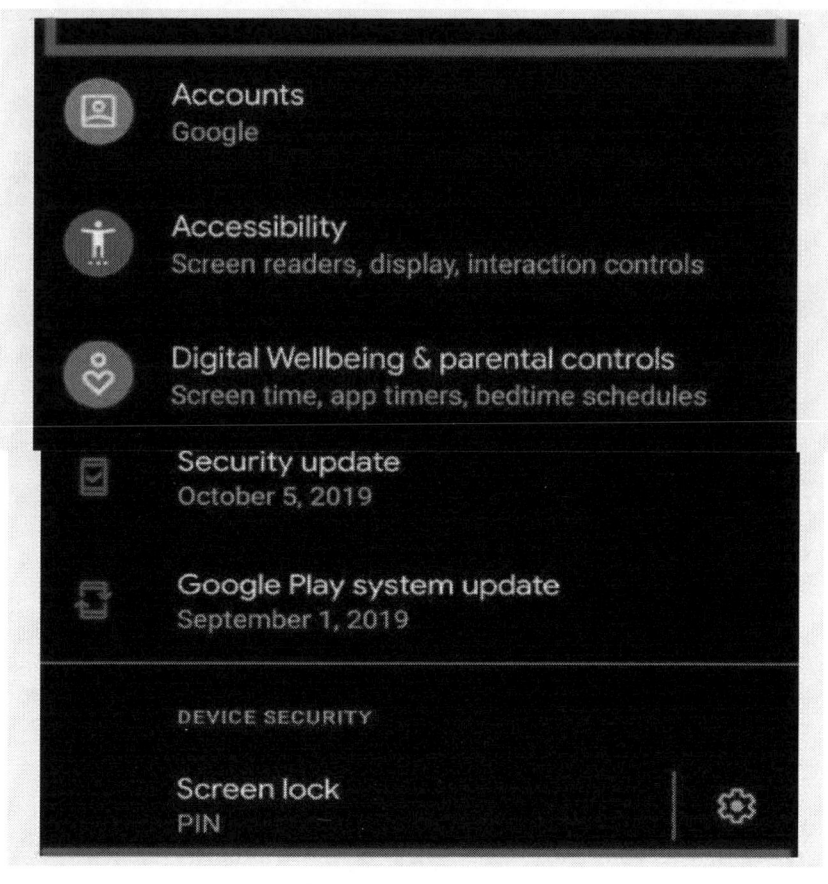

![Settings screen showing Accounts, Accessibility, Digital Wellbeing & parental controls, Security update, Google Play system update, and Screen lock options]

The Face Unlock feature of the Pixel 4 provides you with the option to get your phone unlocked and remain on the lock screen until you swipe up after successful authentication.

- The option of getting your phone unlocked without having to go to the home screen is a wonderful addition because the phone will now be able to hide your notification content until it recognizes the face trying to access it.

108

- All you need to do is swipe the screen to access the phone and it will automatically bypass the lock screen by default heading straight to the home screen.

Setup face unlock
1. Go to Settings.
2. Select Security.
3. Select Face unlock so you can access it in case the method of security was not chosen when setting up your phone.

You can also delete your face data here if you don't need it anymore.

Skip the lock screen

You can choose to make use of the screen whenever you look at your phone to get it unlocked; it will still be able to return to where you left off. However, you might need to lock your screen for privacy reasons by taking the following steps:

1. Go to Settings.
2. Select Security.
3. Select Skip lock screen.

Home Screen

Enable or disable home screen rotation

1. Go to the Home settings.
2. Tap Allow home screen rotation.

Enabling this feature will allow you to view your screen in the landscape portrait as well.

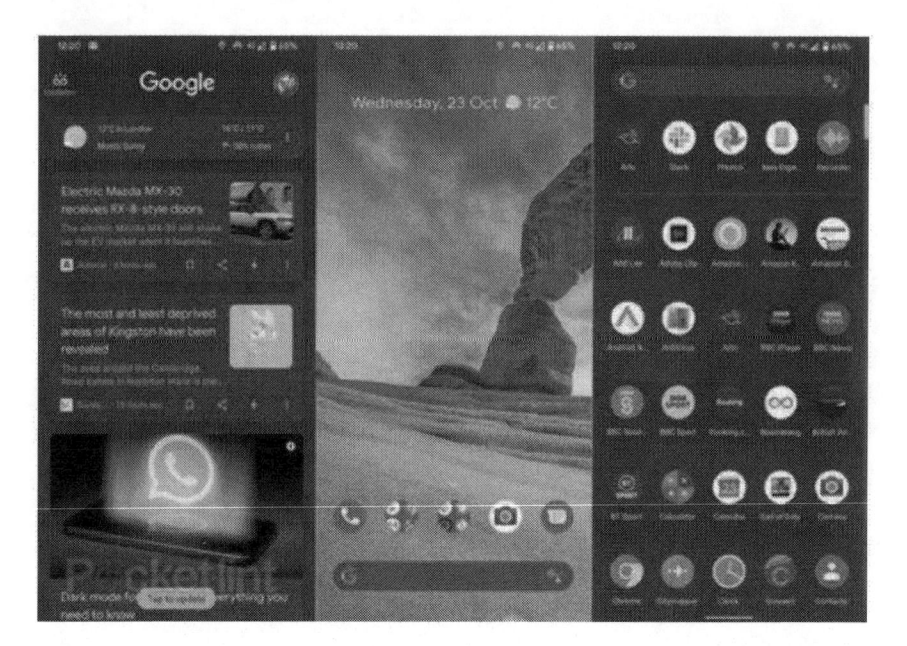

Use touch-free gestures with Motion Sense

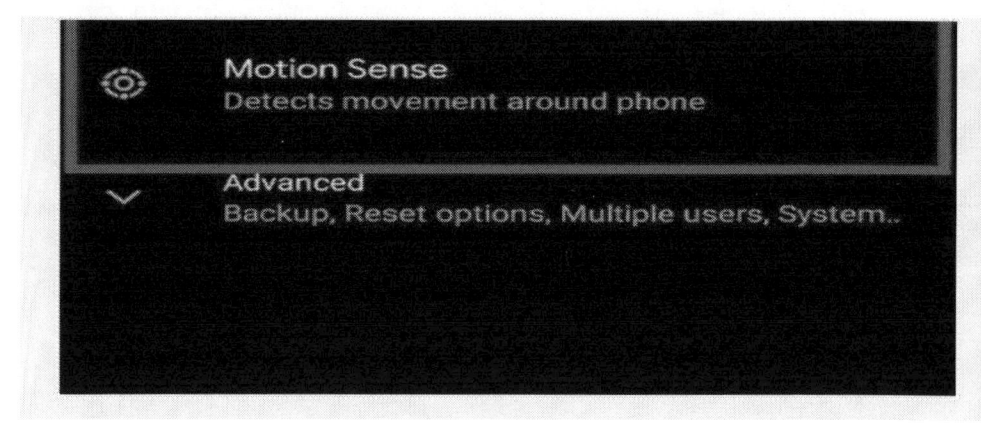

- The Motion Sense feature present in Pixel 4 has included radar technology allowing users to use enabled gesture-based interactions without having to pick up their phone or even touch it at all.
- You can snooze an alarm by waving over your phone and skip songs, silence calls and even check your notifications.

Turn on or off Motion Sense

Your phone can detect your hand using the new radar system and perform actions without having to touch your phone.

1. Go to Settings.
2. Tap System.
3. Tap Motion Sense to turn it on or off.

How to turn on-body detection on or off

i. Go to the Smart lock menu.

ii. Tap on the On-body detection.

iii. Turn the Smart lock On-body detection on or off.

How to Use the On-Body Detection

• As long as your body senses that it's on your body, it will stay unlocked after you must have unlocked your device but once it is placed down like setting it on a table, after about a minute, it will get locked automatically.

• Your device will get locked automatically within 5 to 10 minutes after getting into a car, train, bus or any other vehicle.

It is to be noted that in a situation where the device does not lock automatically, you can go ahead and lock it manually.

• The on-body detection will learn your walk pattern when used with some devices. Due to this feature, the moment it senses a walk that is different from your walk pattern, your phone will get locked automatically.

When this happens and the phone is actually on you, you can simply unlock, then the phone will automatically learn the changes in your walk.

It should be noted that the device makes use of the accelerometer data in storing your walk pattern when you are carrying your device.

The moment the on-body detection feature is turned off, this data will get deleted automatically.

Keeping Your Device Unlocked When It's At a Trusted Place

- Make sure your phone is set to the current location.

- Trusted places will only work best with Wi-Fi.

- It is important to note that your trusted places are based on estimates.

- The trusted places can go beyond the walls of your home or custom places and can keep your device unlocked within the radius of up to 80 meters.

- You should also note that someone that has access to specialized equipment can gain access to unlock your device and location signals can be copied as well as manipulated.

How to show lockdown option in your power settings

i. Open the Settings app ⚙ on your device.

ii. Tap on the Security & location

iii. Select the Lock screen preferences.

iv. Get the Slow Lockdown option turned on.

v. Press the power button so that the lockdown button will show in the list of the settings.

How to Turn On Lockdown

It is good to know that the lockdown will not only work until your phone is unlocked; you can keep using the lockdown anytime you want to by turning it on.

To have the lockdown turned on, you are required to take the following steps:

i. Hold down the power button for some seconds.

ii. Tap on the Lockdown.

How to Add your home location

i. Go to the Smart Lock menu.

ii. Select "Trusted places."

iii. Select "Home."

iv. You will be prompted with on-screen instructions to follow.

How to Remove your home location

i. Go to the Smart Lock menu.

ii. Select "Trusted places."

iii. Select "Home."

iv. Tap on the "Turn off this location" to turn off your home as a trusted location.

v. Tap on Edit and Clear to have your home address cleared from Google.

How to have your home location edited

i. Go to the Smart Lock menu.

ii. Select "Trusted places."

iii. Select the home location.

iv. Select Edit your home address or map.

v. Enter the new address to be used as your home.

In a situation where you have multiple buildings on the same address, the trusted places will recognize the address. You are advised to add the actual home's location within the building complex for better location accuracy.

How to add a custom place

i. Go to the Smart Lock menu.

ii. Tap on the "Trusted places" option.

iii. Tap on the "Add trusted place," this will display a map showing your current location.

iv. Tap on Select this location to make use of your current location.

v. You can pick other locations by tapping on "Search."

vi. Give the trusted location a name.

vii. Tap on "OK."

How to edit or remove custom place

i. Go to the Smart Lock menu.

ii. Tap on the "Trusted places" option.

iii. The list of trusted Smart Lock options will be displayed.

iv. Select your preferred place.

v. Then you have to options to perform one of the following actions:

- Rename

- Delete

- Edit Address

How to add or remove a trusted device

i. Put your Bluetooth on, on your device.

ii. Go to the Smart Lock menu.

iii. Tap on "Trusted devices."

iv. Tap on "Add trusted device."

v. Tap on the device from the list of devices that is displayed.

How to remove a Bluetooth device

i.   Select the device you want to remove.

ii. Tap on the device.

iii. Tap on "OK."

NOTE: Ensure staying secure when using trusted Bluetooth devices

•   Your phone can be unlocked by intruders imitating your Bluetooth connection.

•   In a case, a notification comes up on your phone and the device can't verify if you are on a secure connection, you will get notified. Once you get a notification, you will need to have your phone unlocked.

•   It should be noted that the Bluetooth connection range varies depending on the Bluetooth device, device model and your present environment. You can have a Bluetooth connection range of up to 100 meters.

•   With your trusted device unlocked, anyone that takes your phone nearby your trusted device will have access to your phone.

How to set up, improve or remove a trusted faces

- Once your device recognizes your face, you can get the device unlocked.

- Once the trusted face has been set up, your device will always get unlocked every time you turn on your device and it looks your face.

- It should be noted that this facial recognition is less secure unlike using your PIN, password or pattern because even someone that has a similar face like you can get to unlock your phone.

How to set up and use a trusted faced

i. Go to the Smart Lock menu.

ii. Tap on the "Trusted face☺" option.

iii. Tap on the "Set up."

iv. Finally, follow the on-screen instructions that come up.

Once the device looks your face, through the facial recognition, the trusted face will be displayed and in a situation where the device doesn't recognize your face, unlock manually by making use of your PIN, password or pattern.

How to Improve the Facial Recognition of Your Device

i. Go to the Smart Lock menu.

ii. Tap the "Trusted face."

iii. Tap on the "Set up."

iv. Finally, follow the on-screen instructions that come up.

How to Remove A Trusted Face

i. Go to the Smart Lock menu.

ii. Tap on the "Trusted face" option.

iii. Tap on the "Remove trust face."

iv. The device will verify if you want to remove the trusted face.

v. Tap on the "Remove" option.

• It should be noted that the Smart Lock is not responsible for storing your photos.

• The Google servers does not back up the data used in recognizing your face neither can you see the data, the data is only kept on your device apps.

Chapter 13: How to Use the "Ok Google"

Once you say "Ok Google" to your device while the screen is locked, if your voice is recognized by the device, you can command the Google to perform actions such as visiting sites without having to manually unlock your device.

How to access the Google Assistance using your Voice

- The Voice match can be used on your phone or instruct the Google assistance to perform actions for you by saying "Ok Google" or "Hey Google."
- It should be noted that you must be using a speaker or Smart Display with the Google Assistance built-in.

How to Have the Google Assistant Turned On or Off Android

i. Touch and hold the Home button on your Android phone or tablet or use the Google assistant by saying Ok Google" or "Hey Google."

ii. Scroll to the bottom right.

iii. Tap on the icon ⊘.

iv. Go to the top right of the screen.

v. Tap on your "Profile picture"

vi. Tap on "Settings."

vii. Tap on the "Assistant."

viii. Go to the "Assistant devices."

ix. Select your phone or tablet.

x. Have the Google Assistant turned on or off.

How to Let Your Voice Open the Google Assistant on Your Android Phone

i. Touch and hold the Home button on your Android phone or tablet or use the Google assistant by saying Ok Google" or "Hey Google."

ii. Scroll down to the bottom right.

iii. Tap on the icon ⊘.

iv. Go to the top right.

v. Select your profile picture.

vi. Tap on the Settings

vii. Tap on the Assistant.

viii. Get the Access turned on with the use of your Voice Match.

It is to be noted that even if you turn your Access off using your Voice Match:

• You can still make use of the Google Assistant by touching and holding your Phone's Home screen button.

• You can get your phone squeezed to talk to the Google Assistant on the Pixel 4

How to Teach the Google Assistant to Recognize Your Voice

i. Touch and hold the Home button on your Android 5.0 phones upward or use the Google assistant by saying Ok Google" or "Hey Google."

ii. Scroll down to the bottom right.

iii. Tap on the icon ⊘.

iv. Go to the top right.

v. Select your profile picture.

vi. Tap on the Settings

vii. Tap on the Assistant.

viii. Go to the Assistance devices.

ix. Select and tap on your phone or tablet.

x.   Ensure that your Google Assistance is on.

xi.   Tap on the "Voice Model" option.

xii.   Then tap on Retain voice model.

xiii.   You will be prompted with some instructions to have your voice recorded.

How to Let the "Ok Google" And "Hey Google" Get Your Phone or Tablet Unlocked

•   As long as your voice is recognized by the Google Assistant, your device can be unlocked using the "Ok Google" and "Hey Google."

•   Nevertheless, using this setting can make your device less secure because a similar voice can get to have the device unlocked.

•   It should also be noted that you can't use the "Ok Google" in unlocking the Android 8.0 versions upward but can make use of the Google Assistant to lock the screen

To let the "Ok Google" and "Hey Google" unlock your phone or tablet, you are required to take the following steps:

i. Touch and hold the Home button on your Android 5.0 phones upward or use the Google assistant by saying Ok Google" or "Hey Google."

ii. Scroll down to the bottom right.

iii. Tap on the icon ⊘.

iv. Go to the top right.

v. Select your profile picture.

vi. Tap on the Settings

vii. Tap on the Assistant.

viii. Go to the Assistance devices.

ix. Select and tap on your phone or tablet.

x. Ensure that your Google Assistance is on.

xi. Also ensure that the Access with Voice match is turned on.

xii. To unlock using the Voice Match, turn on.

How to get your voice linked with the Google Assistant device using the Voice Match

• The Google Assistant can be taught to recognize your voice making use of the Voice Match.

- Also in order to make use of your Voice commands so as to get personal results, your voice can be linked to the Smart Display, speaker or the Smart clock.

- About six people's voice can also be linked up with a single speaker using the Voice Match.

- Everyone whose voice is linked up to the device must complete the upgrade so as to have the new version of the Voice Match upgraded.

- It is important to note that these features are not applicable to the Bose or Sonos speakers.

- It is also important to note that the language that can be used depends on the device used.

How to have your voice linked

A Google Account must be linked to the Google Assistant device to make use of the Voice Match and in a situation where multiple Google accounts are involved, you can select the Account you like to make use of.

To link your voice you are required to take the following steps:

i. Get the Google Home app opened.

ii. Scroll to the bottom of the screen.

iii. Tap on the Home icon 🏠.

iv. Tap on your device.

v. Tap on the Device settings icon ⚙.

vi. Tap on the Voice Match.

vii. Tap on the Add icon ➕.

viii. The follow the instructions that you are prompted with.

You are advised to make use of the US English when linking your voice with the Google Assistant to make it easy for the Google Assistant in automatically acknowledges your voice when something is politely asked.

Enable Continued Conversation on Google Assistant

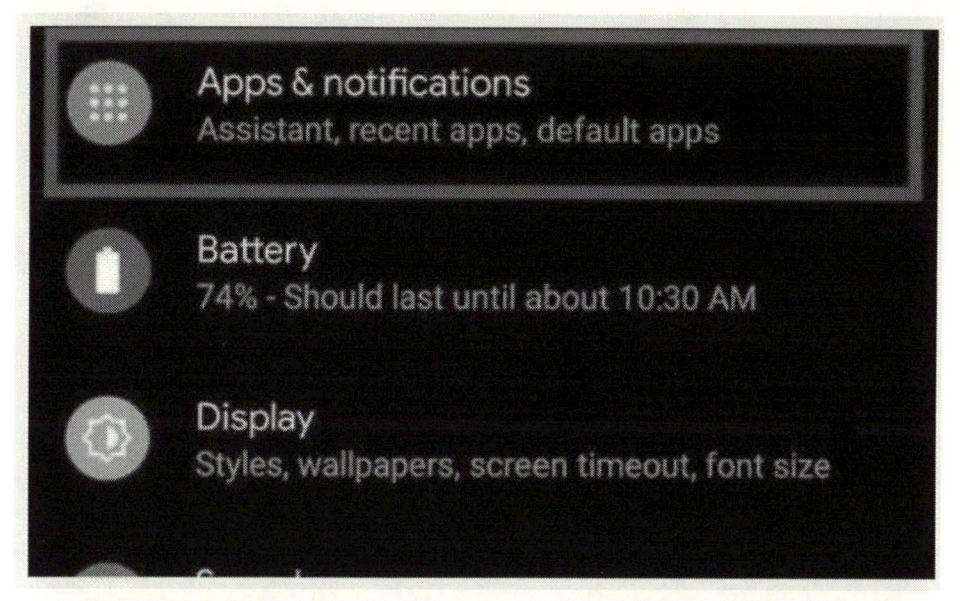

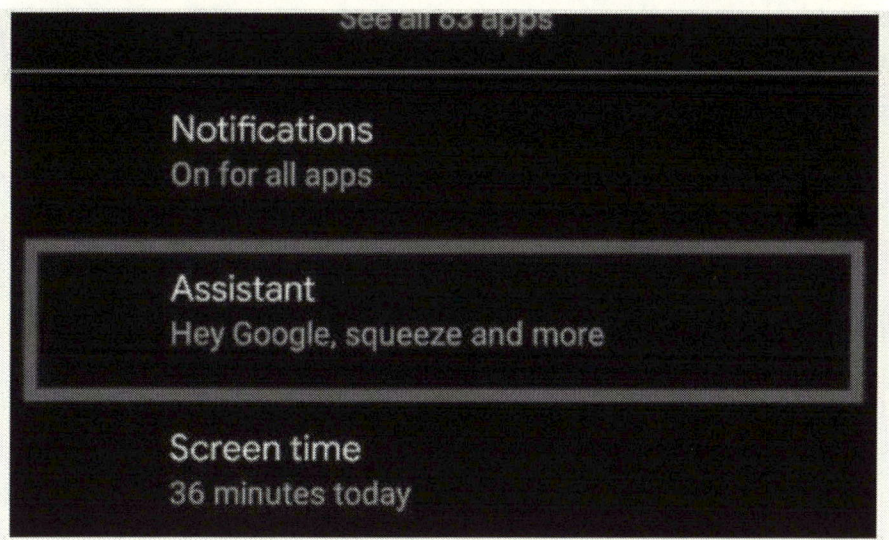

Apps & notifications
Assistant, recent apps, default apps

Battery
74% - Should last until about 10:30 AM

Display
Styles, wallpapers, screen timeout, font size

Notifications
On for all apps

Assistant
Hey Google, squeeze and more

Screen time
36 minutes today

One of the best features that help in showcasing the Google Assistant is the Continued Conversation.

Once it has been enabled, the feature ensures that the assistant keeps listening for further speech after you have made a query or command.

The Assistant is now being introduced to most Google's devices and also getting its feature set as well as its powers expanded with the recent boom in machine learning and Artificial intelligence. This chapter puts you through on things you can try with the Google Assistant.

Squeeze to launch Google Assistant

1. Go to Settings.

2. Tap System.

Tap Gestures.

You will now be able to control Active Edge, configure the squeeze sensitivity or get it disabled if you don't need it. When the screen is off, you can opt to use it.

You can start the Google Assistant listening by squeezing and start to talk.

How to Launch Google Assistant

The new gesture has been introduced to help you in accessing the Google Assistant easily on the Android 10 present in Pixel 4.

All you need to do is to swipe from the bottom corner to launch the Google Assistant or alternatively go to the search bar to tap the icon on the home screen.

Swipe up Google Assistant to Access More Personal Information

After launching the Google Assistant, swipe up to find more information updates that might be waiting for you. You will be able to see what's coming up or check on your commute.

Turn on the OK Google hot word

You will be prompted to set up the Ok Google hot word when setting up your phone initially. It can be set up manually as well in case you didn't set it up initially.

Do this by unlocking your phone and say OK Google and you will be prompted with the setup page.

1. Open an app with Google Assistant

Open Netflix or any other app by saying "Ok Google, open Netflix."

They are also smart like some other apps making it possible for the assistant to navigate contents within them such as watching specific shows on Netflix or searching for a specific artist on Spotify.

I'm feeling lucky

In a case where you are searching for the Google Assistant's Easter Egg, say "I'm feeling lucky." You will be taken to a trivia quiz with loads of fun.

How to See Which Devices Are Linked To Your Voice

i. Get the Google Home app ⌂ opened.

ii. Scroll to the bottom of the screen.

iii. Tap on your account ⊙ icon.

iv. Tap on the Settings option.

v. Tap on the Assistant option.

vi. Tap on the Voice Match.

vii. A list of Smart Displays, speakers and Smart Clocks linked to your device will be displayed.

How to Unlink Your Voice From Your Device

• Unlink your voice so as to stop getting personal results on your speaker, Smart Display or your Smart clock.

• You will need to get your Google Account unlinked so as to have your voice unlinked.

- Once you unlink, you will need to set up your speaker, smart clock or the smart display again using your Google account to continue the use of this feature with your device.

You are required to take the following steps so as to unlink your voice from your device:

i. Get the Google Home app ⌂ opened.

ii. Scroll to the bottom of the screen.

iii. Tap on your account ⊙ icon.

iv. Tap on the Settings option.

v. Tap on the Assistant option.

vi. Tap on the Voice Match.

vii. Tap on the Clear icon and then Unlink to get a specified device unlinked.

viii. Select the "Remove from all devices" option and then tap on the Unlink to have all devices unlinked.

ix. A message will notify you that your Google Account and voice have been unlinked from your device.

Chapter 14: Understanding How Your Voice Works With Other Settings

Upgraded Voice Match

• A unique voice model will be created on the Google servers once the Google Assistant is taught to recognize your voice. This voice will then store on the devices where you have had the Voice Match turned on only.

• The device sends the voice model which the Google assistant has learned to Google to have the voice queried when someone speaks to the device. It will then have it compared to the voice model.

Once processing is concluded, the voice model and comparison information will be deleted automatically by Google.

• Your device will produce the desired personal results once it is able to recognize your voice.

• The query will be treated as a guest query if the voice does not match and you won't get the desired result.

The Voice Match and Personal Results

• If the personal results are enabled on your device by anyone that links their voice with the Voice Match to the Google Assistant when setting up the device, the personal results will become accessible to anyone.

• Once your voice is linked with the Voice Match, if the Google Assistant does not recognize a voice, it will not respond with your personal assistant to the voice.

• A voice that is similar to that of yours can get the personal results, so if not convenient with this you are advised to stop allowing personal results.

The Voice Match and Media services

• Choose specific music and video services when linking your voice with the Voice match. It is to be noted that personalized music and video suggestions can be provided by the Google Assistant.

• The use of your Google Assistant device, which has already been linked to your voice, by other people, may result in the change of your media history as well as recommendations from the services.

- Ask other people to have their own music and video services linked so as to stop them from using your media services.

Get more Pokemon fun

You can download the Pokemon Wave Hello app which is powered with the Motion Sense, letting you interact with a range of Pokemon. As a standalone app, it's like the wallpaper.

Enable Continued Conversation on Google Assistant

One of the best features that help in showcasing the Google Assistant is the Continued Conversation. Once it has been enabled, the feature ensures that the assistant keeps listening for further speech after you have made a query or command.

The Assistant is now being introduced to most Google's devices and also getting its feature set as well as its powers expanded with the recent boom in machine learning and Artificial intelligence. This chapter puts you through on things you can try with the Google Assistant.

Squeeze to launch Google Assistant

1. Go to Settings.

2. Tap System.

3. Tap Gestures.

You will now be able to control Active Edge, configure the squeeze sensitivity or get it disabled if you don't need it. When the screen is off, you can opt to use it.

You can start the Google Assistant listening by squeezing and start to talk.

Chapter 15: How to Add Apps, Shortcuts, and Widgets to the Home Screen

Customize your home screen to get quick access to your favorite contents. The contents that can be added and organized on the home screen includes the apps, shortcuts to the content inside apps as well as the widget which displays information without having to open the apps.

How to add an app to the Home screen

You are required to take the following steps to add apps:

i. Open all apps by swiping up from the bottom of your screen to the top.

ii. Select the App you want to open. In a situation where you are using a Google account, all your apps can be accessed from the work tab.

iii. Alternatively, you can open apps making use of shortcuts by touching and holding the app and choosing an action from the options that come up.

iv. Select and drag the app.

v. Drag the app to your preferred position and release your finger.

How to add a shortcut to the home screen

i. Select the app and hold it before releasing your finger.

You will be prompted by a list if this app has shortcuts.

ii. Select and hold the shortcut.

iii. Drag the shortcut to your preferred position and release your finger.

You can tap on the shortcuts to make use of it without adding it to the home screen.

Enable and control app notification dots

This allows you to be notified by placing dots on your apps whenever you get unread notifications.

1. Long press on the wallpaper.

2. Select Home settings.

3. Select Notification dots.

You will also find a list of your recent notifications here which will provide you with the chance of customizing whether notification dots should be shown or not. However, you can switch this feature off if you don't want it.

Use app shortcuts

You can access some apps on Android 10 using shortcuts instead of having to press and hold their icons. This feature can be used to do the following:

• Capturing videos or photos using a camera

• Navigating home using the Maps.

• Adding contacts

• And many more

All you need to do is press and hold the icon and it will pop up. The app notifications can be viewed directly by using this method.

Create shortcut icons

You can drag and place the list of app shortcuts on the home screen as their icons whenever they pop up on the screen. For instance, you can go to the camera to drag out a shortcut for accessing the selfie camera instantly.

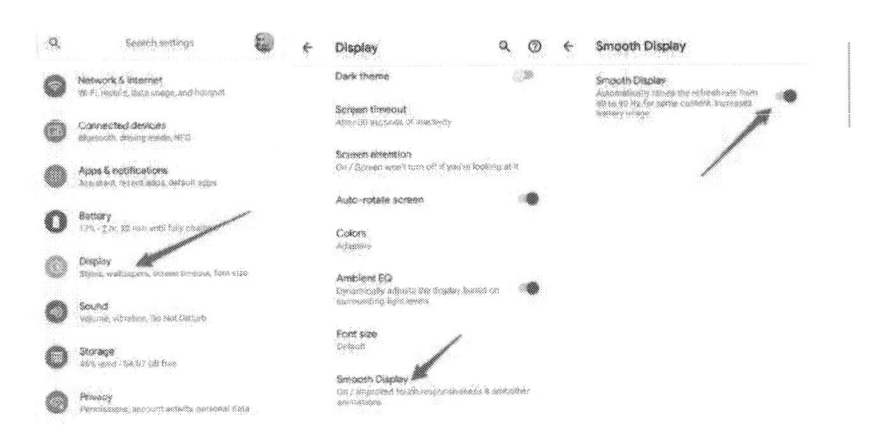

Safety app

- Pixel 4 users can now find the new preinstalled safety app in the app drawer. It helps in emergencies.
- Once you have signed into the app using your Google account, you will be prompted to choose an emergency contact and input any important medical information such as your blood type or if you are having any allergies.
- The most amazing thing about this feature is that the automatic crash detection can be enabled to help alert

emergency personnel. Or your close contacts in case you are involved in an accident.

How To Add A Widget

i. Go to the Home screen.

ii. Select and hold an empty space.

iii. Select the Widgets icon.

iv. Select and hold the Widget.

v. Drag the Widget to your preferred location.

vi. Release your finger to finalize.

It is important to note that the Widget is included in some apps and the Widgets can be accessed by touching and holding the app.

How to resize a Widget

i. Go to the Home screen.

ii. Select and hold the Widget.

iii. Release your finger.

iv. Once you are prompted with an outline with dots on the sides, then the widget can be resized.

v.   Drag the dots to resize the widget.

vi. After resizing, tap the space outside the widget.

Chapter 16: Home Screens Organization

How to make a folder

i. Select and hold a shortcut or app.

ii. Tap and drag the shortcut or app and place it on another one.

iii. Release your finger to initiate your action.

iv. You can touch more apps and drag them on the group to add other apps to the group.

How to name your folder.

i.   Select the group.

ii. Select the "Unnamed folder."

iii. Input the name you like to use.

How to move apps, widgets, shortcuts, and groups

i. Select and drag the item.

ii. You will be prompted with images of your Home screens.

iii. Drag the selected item to your preferred position.

iv. Release your finger.

It is to be noted that the "At A Glance" information at the top of the screen can't be moved on Pixel phones.

How to remove an app, shortcut, widget or group

i. Select and hold the item.

ii. Drag the items and place it on the Remove icon.

iii. Release your finger.

iv. You will be prompted with an option to remove, uninstall or both of it.

The Remove option will take off an app from the Home screen while selecting an Uninstall option will result in totally taking off an app of your device.

How to add a Home screen

i. Select and hold the app, shortcut or group.

ii. Drag it to the right until an empty home screen appears.

iii. Release your finger to finalize adding the app to the empty home screen.

How to Switch Between Recent Apps

i. Go to the Settings app.

ii. Select "System"

iii. Select the "Gestures option."

iv. Swipe up to the middle of the screen from the bottom.

v. To make a switch to the app you like to open, swipe from the left to the right.

vi. Select the app you like to open.

vii. You can also slide the home button to the right to switch between apps.

How to close apps

• To close just one app, all you have to do is to swipe up from the bottom of the screen and swipe up on the app.

• To close all apps, you will need to swipe from the bottom of the screen after which you can scroll to the left and select the "Clear all" option.

• Tap on home to navigate back to the Home screen.

View two apps simultaneously on the Pixel Phone

You can view more than one app at once on your Pixel phone. Though this is dependent on the apps you are using, you can get to do other things while watching a video through a small window.

How to make use of two apps at once (Split screen)

You will need to turn on the "Swipe up on Home button," To turn on this feature, you will need to visit the Settings app, then navigate through the System to Genres and tap on the Swipe up on Home button.

Once you have turned on the "Swipe up on Home button" feature, you can then take the following steps:

i. Launch an app.

ii. Swipe up from the bottom of the screen to the middle.

iii. Select and hold the icon of the app.

iv. Select "Split screen."

v. Two screens will come up.

vi. Select another app on the second screen.

vii. You can also drag the bar between the apps to the top or bottom of your screen to view one app again.

How to view one app while using another app (Picture-in-picture)

Some apps such as YouTube or Google Maps can be viewed in a small window while using other apps. An example is watching a video and doing something else simultaneously. To do this, you are required to take the following steps:

i. Launch an app and play a video simultaneously on your device.

ii. Select the home icon.

iii. Select the Picture-in-picture icon.

iv. Tap and drag the video off the bottom of the screen to close the video.

To maximize the video again, you can tap the video and navigate to the full screen.

How to turn off Picture-in-picture for an app

i. Open the Settings app of the device.

ii. Select "Apps & notifications."

iii. Select "Advanced."

iv. Select "Special app access."

v. Select "Picture-in-picture."

vi. Select an app

vii. Turn off the "Allow picture-in-picture."

Chapter 17: Managing Screen and the Display Settings

Enabling the 90Hz refresh rate everywhere

The new Pixel phones are designed with the 90Hz refresh rate display which is the best panel seen on the Pixel phone to date. The only minor issue this feature has is the adaptive frame rate.

However, you can make use of the Developer options toggle to force the refresh to remain at 90 Hz. Once you have enabled the Developer Options, take the following steps:

Go to Settings.

Select Developer options.

Select Force 90 Hz refresh rate.

Toggle on.

Enable Developer Settings

Enable The Developer Options By Taking The Following Steps:

1. Go To The Settings.

2. Tap About Phone.

3. Scroll To The Bottom And Tap On The Build Number Repeatedly.

4. You Will Be Asked To Confirm Your PIN After Several Taps Which Will Unlock Your Developer Options.

5. Then Go To The Settings.

6. Tap System.

7. Select Advanced.

8. Select Developer Options.

Turn Off The Developer Options

Once Those Options Have Been Unlocked, You Will Be Prompted With A New Section In The Settings Menus As Shown Above. Open It To Access The Toggle Switch At The Top From Where It Can Be Turned Off And The Menu Option Vanishes.

Ambient EQ

This feature can be used to adjust the color and brightness of your Pixel 4 display based on the ambient lighting conditions just like we have in iPhones. Take the following steps to enable the Ambient EQ:

1. Go to Settings.

2. Tap Display.

3. Tap Ambient EQ.

4. Toggle on.

Screen Attention

Screen attention helps to keep the screen of your phone turned on while you look at it by detecting if your eyes are looking at the display using the front-facing camera. You can get this feature enabled by taking the following steps:

1. Go to Settings.

2. Tap Display.

3. Tap Screen attention.

4. Toggle on.

Reach to Check

Reach to check is a newly introduced feature that now helps to highlight the Motion sense. You can enable this feature by taking the following steps:

1. Go to Settings.

2. Tap System.

3. Tap Motion Sense.

4. Toggle on.

5. Toggle Reach to check on.

Adjust the Display's Refresh Rate

The maximum refresh rate of 90 Hz possessed by the display of the Google Pixel 4 makes it useful for routine tasks such as scrolling.

The Pixel screen is designed such that it can adjust its frame rate automatically, switching between the more standard 60Hz and 90Hz. Although, it consumes more battery power in the process, fortunately, the variable refresh rate can be disabled and the display can be limited to 60Hz which is much more standard.

Take the following steps:

1. Go to Settings.

2. Tap Display.

3. Tap Advanced.

4. Select Smooth Display.

5. Slide the switch to the OFF position.

On the other hand, you can force the display of your device to always run at 90Hz from the developer options. Once you have enabled the developer options, take the following steps:

Launch the Settings app.

Go to About Phone.

Go to the bottom of the screen and tap on Build Number until you are prompted with a message that congratulates you on becoming a developer.

Go back to Settings.

Select System.

Select Developer options.

Find the option labeled "Force 90Hz refresh rate" by scrolling down and enable it.

It is, however, important to note that this feature will consume lots of battery power.

Turn on or off the smooth display

You can go to the settings to toggle the 90Hz display on or off.

Google will automatically make use of the adaptive system in moving from 60 to 90Hz in some apps. However, it consumes lots of battery power so you might need to get it disabled if you want by taking the following steps:

1. Go to Settings.

2. Tap Display.

3. Tap Advanced.

4. Select smooth display.

5. Turn off.

6. Turn on the always-on display

7. Go to Settings.

8. Tap Display.

9. Tap Advanced.

10. Select Lock screen display.

You will be provided with the always-on display option which will display the time, date, including the current weather on your lock screen. However, you can save more battery life by turning it off.

1. Turn on tap to wake

2. Go to Settings.

3. Tap Display.

4. Tap Advanced.

5. Select lock screen display.

6. Toggle "tap to check phone" on.

This can be used as an alternative to the always-on display option letting you access the details with a tap.

Get notifications when you lock your phone

1. Go to Settings.

2. Tap Display.

3. Tap Advanced.

4. Select lock screen display.

To access the always-on display by lifting your phone, you can simply turn on the option that enables you to access your time and notification icons without having to press any buttons.

Wake the display on receiving new notifications: You will find the option to wake the display up when new notifications arrive in the lock screen settings.

This will prevent you from getting overwhelmed with notifications apart from draining your battery life faster.

Quickly access the battery details

1. Swipe down to access the Quick Settings.

2. Press and hold the battery saver toggle.

Once you follow these steps correctly, you will be taken to the battery details page.

To see what's consuming your battery

You can quickly access the apps that are eating most of your battery life:

1. Go to the battery panel.

2. Tap on the menu top right.

3. Select Battery usage.

This will present you with a breakdown of the apps eating up your battery life.

Turn on battery saver

1. Go to the battery panel.

2. Tap on the menu top right.

3. Select the battery saver.

You can configure it to switch on automatically when your battery hits 5 percent or 15 percent mark.na

Manage the colors of the display

1. Go to Settings.

2. Tap Display.

3. Tap Colours.

You will be provided with the following options: Natural, boosted or adaptive. Adaptive is the best for most situations.

Night light automatically turns on/off at dusk and dawn:

This feature will help to reduce the blue light from the display to provide you with a better viewing experience at night by reducing the screen brightness and preventing it from straining the eyes.

1. Go to Settings.

2. Tap Display.

3. Tap Night Light.

This will provide you with all the controls needed. You can customize when the Night Light comes up in the schedule option by using the automatic sunset to sunrise.

The brightness of the screen, display size, font size, and rotation settings amongst other things can be managed to your preferred choice.

How to change the display settings

i. Launch the Setting's app of your device.

ii. Select Display.

iii. Select the Settings you like to change.

iv. Tap "Advanced," to have access to more settings

How to Use the Display Settings: Brightness Settings

• Tap on the Brightness level and move the slider, to set the brightness of your screen.

• Turning on the Adaptive brightness will automatically blend the brightness of your screen to the light around.

• It is to be noted that the Adaptive Brightness is always on by default.

• With the Adaptive brightness on, you can change the screen brightness.

Changing the Color of Your Screen at Night

The blue color of the screen can be reduced for easily viewing your Pixel phone in dim light. Making use of the Blue light at night makes it hard to fall asleep. Nevertheless, making use of the red or amber light is better used to help with the eyes adjustment to night vision.

How to automatically change your device's screen color at night

i. Go to the Settings app on your device.

ii. Select "Display."

iii. Select "Night Light"

iv. Select Schedule to help pick start and end times.

With no Schedule option available, you can go ahead to select "Turn on" automatically.

To prevent the night mode from turning on automatically: Go to the Schedule option and select "None" or "Never."

How to change the color anytime you want

i. Go to the Settings app on your device.

ii. Select "Display."

iii. Select "Night Light"

iv. Tap on "Turn on now" option.

v. Make adjustment to the color under the "Intensity" option.

Chapter 18: How to Open Quick Settings

These settings will come in handy to help you have complete control of your phone. Little changes have been made in the Android 10, but this chapter will highlight some of the important tips you will need to master their use.

It is to be noted that you can access the Night light at the Quick Setting as well.

The Quick setting gives quick access to common settings on your device

i. Swipe down from the top of the device's screen to view the first Quick Settings.

ii. Swipe down again to have access to all Quick settings.

Manage quick settings icons

The order of the quick settings tiles can be managed by dropping the usual shade down from the top of the screen and hitting the pencil icon bottom to the left to edit.

Quickly select a Wi-Fi network

1. Swipe down to access the Quick Settings.

2. Press and hold the Wi-Fi icon.

This will take you straight to the Wi-Fi settings and you can also get to monitor what's happening on your Wi-Fi.

Quickly manage Bluetooth

1. Swipe down to access the Quick Settings.

2. Press and hold the Bluetooth icon.

In situations where the Bluetooth connection to your car or other devices is failing, you will be able to see instantly here.

Turn on torch/flashlight

You won't need a separate app for this feature; all you need to do is tapping the button in the Quick settings to put on your flash as a torch. Alternatively, you can simply say "Ok Google, turn on torch/flashlight" and it will be turned on.

Cast Your Screen

• You can watch your Android device on the TV by simply swiping down and tapping the cast screen and it will be sent to the Chromecast. If you can find it there, go ahead and get the Cast tile added to the Quick Settings making use of the method mentioned above.

• Note that this feature is not supported by all apps.

How to turn Settings on or off

i. Tap on a setting to turn it on or off.

ii. Touch and hold a setting to get more options for it.

When Settings it turned off, it will be dimmed.

How to add, remove or move a Setting

i. Go to the top of the screen and swipe down twice.

ii. Select "Edit" by scrolling to the bottom of the screen.

iii. Touch and drag settings to where you want it placed.

iv. Drag a Setting up from "Hold and drag to add tiles," to have Setting added.

v. Drag a Setting down to "Drag here to remove," to remove a Setting.

Recognizing Settings

To see the names and description of settings icon, which is located on the top right of your device screen or in your quick settings, you will need to swipe down twice from the top of your device's screen.

It is important to note that your phone will by default hide settings' icons that are turned off to help in saving spaces on the device screen.

How to change language settings

i. Launch the Settings app of your device

ii. Tap "System"

iii. Select "Languages & input"

iv. Select "Languages"

v. Tap Add a language.

vi. Select the language you like to add.

vii. Select and drag the language to the top of the list.

How to remove a language

i. Open the device's settings app.

ii. Go to the top and Tap More.

iii. Select "Remove"

iv. Select a language

v. Tap Delete at the top

Font Size and Display Size

You can decide to change the font size and display size to your specification to make it easier for you to see.

How to change the Font size

You are required to take the following steps to make your font smaller or larger.

i. Open the Settings app of the device.

ii. Select Accessibility.

iii. Select Font size

iv. Adjust the slider to select your font size.

It is to be noted that the settings of the font size do not apply to the Google Chrome app because it has its own text scaling control.

How to change the display size

i. Go to the device Settings app.

ii. Select Accessibility.

iii. Select Display size

iv. Adjust the slider to select display size

It is to be noted that changing the display size might cause some of the apps to change position.

How to Put Emergency Info on Your Lock Screen

Messages and emergency information can be added to your Android phone's lock screen. This will result in anyone seeing your message and emergency info even without unlocking your phone.

To put a message on your lock screen, you are required to take the following steps:

i. Go to the device's Settings app.

ii. Select Security & location

iii. Tap Settings by the "Screen lock."

iv. Select Lock screen message.

v. Input the desired message.

vi. Tap "Save."

How to Show Emergency Info from Your Lock Screen

i. Go to the device's Setting app.

ii. Tap "About phone"

iii. Select "Emergency information"

iv. If the "Emergency information" option is not available, go back and select "Users & accounts" instead and tap "Emergency information" afterward.

v. Input in the info you like to share

vi. To input medical information, tap "Edit information," if you don't have Edit information available, tap info.

vii. For emergency contacts, select "Add contact," if "Add contact" is not available, tap "Contacts."

viii. To clear the information entered, tap "More," and then select "Clear all."

How to See Emergency Information

i. Swipe up on the locked screen.

ii. Select "Emergency"

iii. Select "Emergency information"

iv. Once Emergency information comes up, tap on it again.

Setting Up A Screen Saver

With the use of the screen saver, your Android device gets to show colorful backgrounds, photos, clock when it's charging or docked.

How to turn off your screen saver

i. Go to the Settings app of your device

ii. Tap "Display"

iii. Select "Advance"

iv. Select "Screen saver"

v. Select "When to start"

vi. Select "Never"

How to set your screen saver

A. Choose what your screen saver displays.

i. Go to the Settings app of the device.

ii. Select "Display"

iii. Select "Advanced"

iv. Select "Screen saver" and "Current screen saver"

v. You will be prompted with Options to choose from:

Clock: Here you get to see a digital or analog clock as your screen saver.

Colors: Shows changing colors on your screen.

Photos: Uses photos from your Photo apps.

Other apps: The other apps are the apps downloaded to be used with your screen saver. They will be listed here.

B. Choose when you screen saver is displayed

The screen saver can also be used when the device is charging, docked or in both states.

i. Go to the screen saver setting.

ii. Select "When to start."

iii. Tap "More," if the "When to start" option is unavailable.

iv. Then now tap "When to start"

OR

You can also tap "While docked" or "While charging" or docked

C. Testing your screen saver

Tap the "Start now" option to see what your screen saver looks like

If the "Start now" option is not available, Tap "More and tap "Start now" afterward.

Starting Your Screen Saver from the Sleep Mode

When your device goes to the Sleep mode, you will be able to see your Screen saver if you had it pre-set.

To set how long it takes for your device to go into the sleep mode, you are required to take the following steps:

i. Go to the device's setting app.

ii. Select "Display"

iii. Tap "Advanced"

iv. Tap "Sleep"

v. Select Option.

It is to be noted that if you turned off your device with the power button, the screen saver won't start. You are advised to allow the device to go into the sleep mode itself by leaving your screen on.

Split-Screen Multitasking

The Split-screen multitasking feature is now provided by Android and makes use of the Overview to control it.

1. Swipe up to access the Overview.

2. Go to the top and tap the app icon.

3. This will provide you with the split-screen as an option. Tap on the option.

4. Scroll through Overview to search for the second app or open another app.

To return to single screen/not split

1. Press the Home button anytime you find yourself stuck in the split-screen.

2. Swipe down any app found at the top, if any and it will go back to the full screen.

3. Press the home button again and you will be back to default.

Change the Default App

You can choose your default app for apps that can do the same things by taking the following steps:

1. Go to Settings.

2. Tap Apps & notifications.

3. Tap Advanced.

You will be provided with the default apps option where you can easily set your default browser, SMS app, launcher, etc.

Control App Permissions

You can personally manage all permissions for each of your apps by taking the following steps:

1. Go to Settings.

2. Tap Apps & notifications.

Your recent apps will be displayed at the top where you can click through quickly as well as get to edit permissions for your individual apps.

Disable picture-in-picture

This feature helps to allow a thumbnail version of an app or video to play once it returns to the home screen. It's a good feature but can't be disabled if you don't want it by taking the following steps:

1. Go to Settings.

2. Tap Apps & notifications.

3. Tap Advanced.

4. Select Special app access.

5. Tap Picture-in-picture.

You can toggle the apps you don't like to use here such as toggling off the Chrome which will stop you from getting in-browser videos playing as picture-in-picture.

How to Change the Wallpaper On Your Pixel Phone

i. Go to the Home screen of your phone

ii. Touch and hold the empty space on the display.

iii. Tap "Wallpapers"

iv. Tap My photos to use your personal images.

v. For curated images, tap a category and tap an image.

vi. Tap "Set Wallpaper" at the top of the screen.

vii. Select screens to show wallpaper (if available).

viii. Tap "Set wallpaper" at the top.

ix. Pick screens to show wallpaper, if available.

It is to be noted that whatever wallpaper you choose will determine the color of your notifications, quick settings as well as other apps.

How to Skip Daily Wallpaper

In a situation whereby you are on a plan whereby your wallpaper changes automatically daily, you can take the following steps to skip:

i. Go to the Home screen of your phone.

ii. Touch and hold an empty space on your screen.

iii. Tap "Wallpapers"

iv. Go to your current daily wallpaper image.

v. Tap "Refresh."

Play with your Pokemon wallpaper

Go to the coming alive section to change your wallpaper to Pokemon and you will get to interact with it using the Motion Sense.

Make the Pokemon jump by tapping on them

Express love by showing your palm.

Change to another character by double-tapping.

Get More Pokemon Fun

You can download the Pokemon Wave Hello app which is powered with the Motion Sense, letting you interact with a range of Pokemon. As a standalone app, it's like the wallpaper.

Access the wallpapers options by:

1. Long press the home screen.
2. Select Styles & wallpapers.
3. Go to the bottom of the screen to select the Wallpapers tab if needed.
4. Tap the Come Alive section.

All wallpapers that come up in this section are all interactive while also moving subtly and making use of the Motion Sense feature to respond to movements. It is an improved version of the popular Live Wallpapers in Androids. You will also find the interactive Pokemon Sidekick wallpaper here.

Home screen tips and tricks

The Pixel launcher is specifically designed for Pixel phones to provide users with what Google thinks is the best experience in Android. It serves the Google Assistant and the Google app with information and news they need.

How to Pick A Live Wallpaper

Various kinds of wallpapers designed with active elements moving subtly in them are provided by Pixel thereby adding some movements to your home screen.

1. Long press on the home screen.
2. Select Styles & wallpapers.
3. Go to the "coming alive" section.

You will find different live wallpapers here as well as the interactive Pokemon option.

Access Google app/Discover and customize it

On Android, pages have always been pushed to the left of the home screen. The once called "Google Now" is now referred to as "Discover" containing a digest of topics that will be interesting to you. You can tap on the slider located at the bottom to see more or less of every story you are shown or go to the menu button to block or say you are not in support of the topic or publication.

Turn off Discover/Google app

You can turn off this feature by taking the following steps:

Go to the home settings.

Tap display Google app and you can turn it on or off here.

Chapter 19: How to Take a Screenshot on Your Pixel Phone

A Screenshot is an image picture taking from the screen of your Pixel phone. After a screenshot is taking you can view, edit as well as share the image of the Pixel phone.

To take a screenshot, take the following steps:

i. Open the screen.

ii. Tap and hold the power button of the phone for a few seconds.

iii. Tap "Screenshot"

iv. The picture of the screen will be automatically taken and saved by the device.

v. The Screenshot capture will be displayed at the top of the screen.

In case the Screenshot capture icon doesn't come up, return to the Home screen.

How to view your screenshot

i. From the top of the screen, swipe down with the Screen capture.

ii. A Screenshot saved notification will appear.

iii. Save the Screenshot save notification.

How to see all the screenshots

i. Go to the Photos app on the phone.

ii. Select "Menu

iii. Select "Device folders"

iv. Select "Screenshots"

How to share Screenshots

To share screenshot:

i. View it.

ii. Select Share

How to Edit your device's screenshot

i. Go to your Photos app.

ii. View it.

iii. Select "Edit "

Editing Photos

This enables you to add filters, crop photos that are available on your device.

It is to be noted that your edited photos will automatically sync to your Google Photos library if you have the backup & sync.

On your computer, iPhone, iPad or Android, you will need to get the Google photos app downloaded and installed to get started.

How to adjust, rotate and crop a photo

i. Go to the Google Photos app on your device.

ii. Open the photo.

iii. Tap "Edit "

iv. Tap "Photo Filters " and select Apply a filter.

v. Tap "Photo filter" again to adjust.

vi. Tap Edit to manually change lightning color and add effects.

vii. Tap Crop & rotate, to crop or rotate. You can tap and add the edges to crop.

viii. Go to the top right and select Save to save your edited picture.

Once you save, the changes will be added to your photos and you can undo at any time.

Back up your photos to have your edits saved in the Google photos.

How to Save Shots from Your Motion Photos

The Google photos app is designed to recommend another shot from a motion photo taken by you. You have the chance to pick your favorite shot as well:

i. Open "Google Photos app" on your device.

ii. Open "Motion photo"

iii. Swipe up on the selected photo

iv. Tap Shots

v. Scroll through the shots and select your preferred.

vi. Tap "Save copy," to save a new photo.

The shots taken will be automatically saved in your Google photos next to the original photo.

How to change the date and time stamps

To change the date and time stamps of a photo or video:

i. Visit photos.google.com

ii. Tap "Computer"

iii. Follow the on-screen prompt.

The computer is used to change the date and time of the Photos as well as videos.

Once a change is made to the date and time of your photo, it will automatically appear in your Google photos.

It will show the original date and time the photo was taken with the camera if it is shared with other apps or it is downloaded.

How to undo edits on your photos

i. Open the Google Photos app.

ii. Open the edited photo

iii. Tap Edit.

iv. Select More

v. Select "Undo edits"

vi. Select "Save"

How to Search Your Photos Using People, Things & Places

You can search for your photos using any term such as your best friend, A pet, etc. Nevertheless, it is important to note that some features are not available in some countries.

How to search for photos

i. Go to the Google Photos app.

ii. Log in to your Google Account.

iii. Go to the top and enter a search term for what you are searching for.

A. How to find photos of a person or pet

i. Go to the Google Photos app.

ii. Log in to your Google Account.

iii. Tap the search bar at the top.

iv. You will be prompted with a list of suggested searches having a row of faces.

v. Tap a face to see the photos attached to it.

vi. Tap Next to see more faces.

In some situations, the row of faces won't show up, this is because:

- Feature not available for your country.

- Turned off in the settings

- The device doesn't recognize the face groups.

On the other hand to find the photos of a person without searching:

i. Tap Albums.

ii. Select "People."

B. How to apply a label

i. Go to the top of a face group.

ii. Tap "Add a name."

iii. Input a name or nickname.

Once this is done, you will be able to make use of the search box by searching with that label. Face labels that are kept private can only be viewed by you, even if the photos are shared.

Engage or Disable Search Box Effects

1. Go to the bottom of the screen.

2. Press and hold on the search box.

You will find the option to enable or disable special effects here which updates your Google with the Google Doodle whenever it celebrates.

How to Turn the Face Grouping Feature Off Or On

You can stop your photos being grouped as similar faces together by turning the "Face grouping" off. Once this is turned off:

- Face groups available in your account will be deleted

- Face models used in creating those face groups will be deleted.

- Created Face labels will be deleted

How to turn off Face grouping

i. Go to the Google Photos app.

ii. Log in to your Google Account.

iii. Select "Menu"

iv. Select Settings.

v. Select Group similar faces feature.

vi. Have the "Face grouping" feature turned off or on.

vii. Turn off Show pets with people for only pets.

How to Combine Face Groups

In a situation where we have the same person appearing in more than one group, they can be merged. You can do this by:

i. Use a name or nickname in labeling one of the face groups.

ii. Use the same name or nickname in labeling the other face group.

iii. Confirm the second name or nickname.

iv. You will be prompted by Google Photos if you would like to merge the face groups.

v. Tap "Yes," if they are the same person.

How to Change or Remove a Label

The face group label can be edited or removed.

To have the label changed:

i. Tap More.

ii. Tap Edit name label.

To remove a label:

i. Tap More

ii. Tap Remove name label

How to Remove a Face Group from Search Page

i. Go to the Google Photos app.

ii. Log in to your Google Account.

iii. Tap the search bar at the top.

iv. Tap next, which is next to the row of faces.

v. Tap more at the top right.

vi. Tap Hide & show people.

vii. Select the faces you like to hide

viii. Tap the Box again to display a face.

ix. Tap done at the top right.

It is to be noted that removing an item from a group doesn't mean the photo or video will be deleted from The Google Photos library.

How the Face Grouping Feature Works

It searches to detect if any photos have a face in them.

The algorithmic models are used to check for similarities occurring in different images once the facing group feature is turned on. It will also analyze to establish if two images are represented by the same face.

Once the check is over, the photos seem to represent the same face are grouped together. You can check manually to see if your images are correctly grouped.

Once you have the face grouping turned on, photos might be included based on their characteristics by the Google photos. Example of such photos is one in which a person is putting on the same clothing across different photos.

Features of the Face groups, face labels, and sharing

• It is to be noted that face groups are not shared when sharing photos.

• Face groups can't be shared across accounts but are private to each account.

• The face groups and labels available in your accounts are only seen by only you by default.

• In order to help Google photos apps in quickly recognizing your face in photos, you can have your face group labeled as "Me". This will, in turn, help your contacts to share your photos with you.

• Turning the face grouping, insinuates you want to make a mode of your face.

• Turn off Face grouping to delete your models

Chapter 20: Camera and Photos

Google has improved greatly with their camera right from the first Pixel that they developed and the Pixel 4 has not been left out as well. They have also added some new features to their camera app design such as the slight UI refresh.

Hold shutter button to record video/Top Shot for short video

You can record a video quickly while in the photo mode by pressing and holding the shutter button. Recording a video this way won't result in a lower-res video, though it will have a resolution of 768 by 1024 pixels.

Nevertheless, the Top Shot can be used for short videos after which you can export a 3 MB HDR image from the resulting clip.

Double-tap to zoom

Using the double-tap available in the viewfinder of your Google Camera app will enable you to zoom in 2x while also revealing the scrollable zoom controls, instead of having to use the pinching gestures. You can zoom in up to 8x by scrubbing the bar forward or backward.

Dual exposure controls

You can easily activate the Dual exposure controls available on the Pixel 4 by simply tapping the viewfinder which will result in the display of the new controls.

The Shadow adjustment can be found below the exposure levels. If you can find the perfect levels for your shots, the adjustments can be locked.

These settings will go back to default once you close the camera or by tapping on the viewfinder again.

Social Sharing

This feature will allow you to share your pictures with family and friends. To use this feature:

Take a Picture

Swipe up on the mini-preview bubble to share an app of your choice instantly.

Google Lens features in Camera

Google has been able to add some core Google Lens features into the Google Camera app.

You will be able to use your camera in doing the following:

Scanning Of Documents

Translation of a small selection of languages

All you need to do is pointing at a document or text and you will be prompted with a small pop-up with which the new features can be activated.

Quick Share in the Camera App

You will notice a small thumbnail having an arrow next to it when you take a picture in the camera app. This is referred to as the Social Share providing you with shortcuts to quickly share the photo you just captured with other social apps.

If you are using this feature for the first time, you will be required to turn on your preferred apps or services in the Social Share drawer. Whenever you tap on the arrow afterward, you will always find the apps that have been added available in the drawer.

Storage Saver

You may find yourself running out of space in no time with the base storage of 64GB available for all the Pixel 4 models. They are however made up of a feature called Storage Saver which helps in managing your phone storage by deleting photos and videos that have already been backed up for some time.

However, you will need to enable Smart Storage by taking the following steps:

1. Go to Settings.

2. Tap Storage.

3. Tap Smart Storage.

You will have the option of picking between waiting 30, 60 or 90 days before the local copy of photos are removed.

New shooting modes have also been added to the Google Pixel 4 apart from the additional lens, giving it an upper hand over the Pixel 3.

Quickly launch the camera

Launch the camera quickly by double-pressing the power/standby button. You will find the settings for this control by taking the following steps:

1. Go to Settings.

2. Tap System.

3. Tap Gestures.

You can turn the "jump to camera" option here thereby providing you with quick access from any screen including the lock screen.

Swipe between photos, videos, other camera modes

Instead of having to hit the buttons, you can switch between photo and video capture and other modes available in the viewfinder by just swiping.

In the landscape, just swipe up or down the screen and from the left to the right in the Portrait mode which enables you to switch from the photo to video capture.

Find the Camera Settings

You will find a drop-down arrow at the top when your device is in the portrait mode and on the left when in landscape mode. Swipe it down and the camera settings will be opened.

It is important to note that this feature is specific to the shooting mode; you can change the video settings by going to the video mode to access it.

Turn Off the Shutter Sound

1. Launch the camera settings.

2. Tap the cog.

This will take you deeper into the settings where the shutter sound can be turned off.

Turn the Frame Hints On Or Off

The Pixel has been designed to provide you with suggestions on how you can take better photos. You can also turn off the feature if you don't need the suggestions by:

1. Launch the camera settings.

2. Tap the cog. This will take you to the deeper settings menu.

3. Toggle the "Framing hints" on or off.

Prioritize your friends in photos

Pixel 4 through the AI can identify people you mostly take pictures with, ensuring they look their best in the picture captured.

You can activate this by going to the deep settings menu where you will find the "frequent faces" option. This will place people close to you as a priority when taking some random pictures.

Customize Your Instant Sharing Options

Once you take a photo, you can be provided with an option to share it instantly through different social platforms. You can do this by taking the following steps:

1. Take a photo.

2. Tap on the arrow that is displayed next to the preview image.

3. Tap the "+" icon.

Then you will be taken to the menu where you will be provided with the sharing options to choose from.

Preserve and share depth data in photos

You have the option of sharing the depth data captured by the camera that enables you to edit in other apps or social platforms.

You can find this option in the menus where you can toggle it on. However, the app warns that it will take longer to process photos when you have this option enabled.

Instant 2x Zoom

The 2x zoom is a newly introduced lens on the rear of the camera. Instead of having to use buttons for this feature, you can activate it by double-tapping , which navigates you to the 2x zoom by switching to the second lens.

Zoom in more

Access the zoom slider by tapping the display to focus option. The zoom slider allows you to zoom up to 8x. Alternatively, you can control zoom by pinching on the display.

Use Night Sight

In dark situations, an on-screen prompt will appear to turn on the Night Sight.

However, you can also turn it on even if it's not dark by simply swiping through the photo modes to find the Night Sight.

Use Astrophotography Mode

This feature stabilizes your Pixel 4 phone even in dark conditions. You will be provided with an on-screen prompt to activate this mode and you will get the final photo after about 4 minutes.

Adjust the Highlights and Shadows

The Pixel allows you to independently change the Highlights and shadows of your photos.

Activate this feature by tapping on the viewfinder to meter the scene and two sliders will pop-up which allows you to change highlights and shadows to get your preferred picture. Once you are satisfied, hit the shutter button.

Lock the Exposure and the Focus

This involves ensuring that the camera locks on the right exposure and focuses on the subject in the frame and sticks to it until the picture is taken.

It enables your camera to keep focus when it has a lot going on.

Enable/disable Motion Photos

You can make your camera capture a short burst video when you snap a photo. This option can be activated in the settings. All you need to do is swipe and it will be enabled all the time or it can be set to auto which makes it take motion photo when it thinks it needs to.

Get Google Lens Suggestions

This feature provides you with certain information through the camera. All you need to do is pointing your camera at a name, phone number or website and you will be prompted with a link to open the chrome browser, make a call or access your contacts with that person.

The feature is enabled by default. Take the following steps to find it in "more":

1. Select Settings.

2. Select Google Lens suggestions.

Engage Google Lens through the Camera

The Google Lens can detect objects and provide you with information through its in-built AI system. It can be located in the "more" option on your camera or press and hold the viewfinder to access it. It will then flip to Lens and search for things for you.

Engage Portrait Mode

1. Swipe to Portrait.

2. Line up your subject.

3. Take the picture.

This feature applies to both front and back cameras.

Engage Beauty Mode

This can also be referred to as face retouching. You will find this feature in the settings, so you are required to swipe open the menu and select from the following options: none, natural or smooth. You can use this feature on both the front and back cameras.

Adjust the Depth Effects in Portraits

1. Go to the Google photos.

2. Open the portrait. A little portrait icon should appear on the image indicating that it can be edited.

3. Tap the edit button.

4. Tap on the edit icon again, and then you will be provided with the sliders for editing.

5. Change effects by sliding blur up or down.

6. Save.

Engage Video Stabilization

If you like to turn on the video stabilization feature, go to the settings menu to access the option.

Extract a frame from a video

1. Go to the gallery to open the video.

2. Select the edit button and you will be provided with the timeline of the video you can scan from.

3. Hit the button to export an HDR image from the frame.

Live Captions

The Pixel 4 now allows you to get an accurate transcription of any video you play on your phone, thanks to the Google AI system.

You will be able to caption your audio messages and videos when you don't have the opportunity to listen.

This feature is handy for the deaf or people who have difficulty with hearing or just can't listen at the moment. You can use this feature across all apps with phone calls and videos as an exception. There is an option to also mask profanity and get sound labeling enabled for things such as laughter.

Presently, it's only compatible with English but more languages will be added soon.

How to activate Live Captions

The Live Captions can be activated in two ways:

1. Go to Settings.

2. Tap Sound.

3. Tap Live Captions.

4. Toggle on Live Captions.

OR

1. Tap the volume button up or down.

2. Tap the small Live Caption logo which has a small box with lines inside.

Hide Profanities

In a situation where you have kids using your phone or you are watching videos in public, you might need to hide profanities.

You can enable by taking the following steps:

1. Go to Settings.

2. Select Live Captions.

3. Toggle Hide Profanity

4. Add sound labels

If you want to make use of the Live Captions, you are advised to add sound labels for music and your background sounds to enhance your experience. You can enable this by taking the following steps:

1. Go to Settings.

2. Tap Sound.

3. Tap Live Captions.

4. Toggle "Show sound labels."

Chapter 21: How to Print from Your Device

Add a printer that uses a working Wi-Fi or mobile network, so as to be able to print from your device. Once this is available, take the following steps:

i. Open the Settings app of your device.

ii. Select "Connected devices."

iii. Select Connection preferences.

iv. Select Printing.

v. Select a print service such as the Cloud print.

vi. Turn "Print service" on or off.

How to view and add printers

i. Open the Settings app of your device.

ii. Select "Connected devices."

iii. Select Connection preferences.

iv. Select Printing.

v. Select Cloud print

Tap More to get more options to add or manage print settings.

How to Add and Use a New Print Service

Adding a new print service:

i. Open the Settings app of your device.

ii. Select "Connected devices."

iii. Select Connection preferences.

iv. Select Printing.

v. Select "Add service"

vi. Input in the printer's information.

Making use of a new printer after it has been added:

i. Open the Settings app of your device.

ii. Select "Connected devices."

iii. Select Connection preferences.

iv. Select Printing.

v. Select the print service

For the management of the print settings, select "More "

It is to be noted that you can't get all apps to print. You can easily take a photo of the screen of the app and print out the photo.

How to Print Gmail Messages

Individual messages, as well as all messages present in a conversation, can be printed.

The first step to take before printing is to make sure your printer is connected using CloudPring or AirPrint.

How to Print a Single Email

You can print out a single email present in a conversation with multiple emails by taking the following steps:

i. Go to your computer.

ii. Open www.gmail.com

iii. Open the email you are printing.

iv. Go to the top right of the email.

v. Tap on the Down arrow.

vi. Select Print.

How to Print an Email That Contains Replies

i. Go to your computer.

ii. Open www.gmail.com

iii. Open and select the conversation you like to print.

iv. Go to the top right.

v. Select print all

It is to be noted that if the sender has turned on Gmail Confidential prior to sending an email, you will be unable to print message texts as well as attachments.

You will be notified with a message which reads "Can't print email or attachments"

How to Print the Calendar

i. Go to your computer.

ii. Open the Google Calendar.

iii. Go to the top right.

iv. Select from the options, Day, Week, Month, Year, or schedule up to the range of date you like to print.

v. Select Settings at the top right.

vi. Tap on Print.

vii. Editing of details such as font size and color settings can be done on the Print Preview page.

viii. Select Print to display a window with print options.

ix. Finally, select Print at the top left.

How to Hide Your Calendars

i. Go to the left side under the "My calendars" or "Other calendars" to choose which calendars you like to hide.

ii. Select the name of each calendar you like to hide.

iii. If the square next to the calendar is outlined, then it is hidden.

iv. Go to the top of the page and follow the steps to print.

You can click a calendar to display the calendar after you are done.

How to Have Your Calendar Fitted On a Page

It is to be noted that only some days will easily fit on a page. To fit your calendar on a page, the following tips will help:

i. Change the Calendar to the Landscape or Portrait mode.

ii. Go to the Print Preview page and select the Orientation menu.

OR

i. Change the calendar view into a different view.

ii. Select Month view if you don't require much detail.

iii. From the Print Preview, edit the date range.

OR

You can hide weekend events to help you in saving more space by taking the following steps:

i. Go to the top right.

ii. Tap on Settings

iii. Go to the View options section and select Show weekends.

OR

Change the Font size by going to the Print Preview and editing the font size to your preferred size.

How to Save As A PDF

A PDF copy of your calendar can be saved on your computer by taking the following steps:

i. Repeat the steps to print your calendar.

ii. Select "Destination" available in the final window of the print options.

iii. Select "Change."

iv. Select Save as PDF under the Print Destinations option.

v. Select Save at the top left.

Go to where the download is saved on the computer to find the PDF of your calendar.

Get calendar and travel details at the top of your home screen

You can easily get entries as well as travel information from Google and send it to your home screen where they can be viewed easily, thanks to the "At a Glance" feature.

1. Go to your home screen.

2. Long press on the wallpaper.

3. Tap "Home screen settings."

You will be able to turn any information you want on. These may include flights, traffic, calendar, etc.

How to Print or Change Page Setup

The page setup can be printed from Google Docs, Sheets as well as Slides apps on your device. To this, you are required to take the following steps:

i. Download Google Docs, Google Slides, or Google Sheets.

ii. Set up your computer for printing by connecting with the Google Cloud Print.

iii. To print from your phone or tablet, open the Google docs, slides or sheets apps that have been downloaded.

iv. Tap More on the file you like to print.

v. Select "Share & export"

vi. Select Print

vii. Follow the on-screen instructions to select the printer.

viii. Tap Print.

How to change the Google Doc page setup

To change the Google doc page set up, the first thing is to have the Google Docs apps downloaded.

How to change the page, size, color or orientation

i. Open the Google Docs app.

ii. Open a document.

iii. Go to the bottom right.

iv. Select "Edit"

v. Tap More at the top right.

vi. Select "Page Setup."

vii. Choose the setting to be changed. Such as the Orientation, Paper size or Page color.

viii. Edit

How to see and change how your document appears when printed

i. Open the Google Docs app.

ii. Open a document.

iii. Tap More at the top right.

iv. Turn on the Print Layout.

v. Select "Edit"

How to use the Google cloud print app for the Android

The Google Cloud print app available on Android can be used to:

i. With the help of the Google cloud print, you can manage your print jobs across multiple accounts.

ii. It can be used to print documents to any cloud-ready printers.

iii. It can be used to have your printer options controlled.

iv. It can be used to view previous print jobs.

How to install Google cloud print

i. The first step is to ensure that the Cloud print is installed on your phone.

ii. Go to the Settings.

iii. Go to the "System" section.

iv. Select Printing.

It is important to note that you will need to download and have the Cloud print app installed from the Google play store if you don't see the Printing option.

How to Share Your Printer with a Specific Person or Group

Anyone that has a Google account can definitely share your printer with you. To share, you are required to take the following steps:

i. Open the Chrome on your computer.

ii. Sign in with a Google account that was used in setting up your printer. If you don't have a Google account you will need to sign up for one

iii. Visit the google.com/ cloudprint#printers

iv. Select your Printer.

v. Select "Share"

vi. Input Gmail address or the Google group you like to share with into the box that comes up.

vii. Change the "Can print" option to the "Can manage" option so as to give another person the chance to rename and share access to your printer.

viii. Tap on "Share."

If you like to share with other groups of people: Add their Google accounts to a Google group.

How to share your printer with everyone

STEP 1: Public sharing should be set up

i. Open the Chrome on your computer.

ii. Sign in with a Google account that was used in setting up your printer. If you don't have a Google account you will need to sign up for one

iii. Visit the google.com/ cloudprint#printers

iv. Select your Printer.

v. Select "Share"

vi. Select Change which is next to the Private option.

vii. Select Anyone with the link to have access to the printer.

viii. Select Save.

STEP 2: Sharing the printer.

i. Copy the link in the "Link to share" box.

ii. You can give the link to whoever you like to share your printer with you.

iii. You have the option to pick how many pages you want to be printed per day.

How to stop sharing a printer

i. Open the Chrome on your computer.

ii. Sign in with a Google account that was used in setting up your printer. If you don't have a Google account you will need to sign up for one

iii. Visit the google.com/ cloudprint#printers

iv. Select your Printer.

v. Select "Share"

vi. Delete a person or group.

How to stop sharing a printer with everyone

i. Open the Chrome on your computer.

ii. Sign in with a Google account that was used in setting up your printer. If you don't have a Google account you will need to sign up for one

iii. Visit the google.com/ cloudprint#printers

iv. Select your Printer.

v. Select "Share"

vi. Click "Change"

vii.   Select Private

viii. Select Save

How to save a file to the web instead of printing

Web documents such as receipts and itineraries can be saved as digital PDFs in the Google drive.

Once they are saved this way, they can be viewed on any device that is connected to the internet and won't need to be printed out.

To save it this way, you are required to take the following steps:

i.   Open Google Chrome.

ii. Tap on "More" at the top right.

iii. Tap "Print"

iv. Select "Change" under Destination.

v.   Go to the search bar and type in "save to Google drive"

vi. Tap on the "Save to Google drive" under the "Google Cloud Print."

vii. Select "Save"

Finding and deleting files

How to find and open files

i. Go to the Files app, to display the downloaded files.

ii. Select "Menu" to find other files.

iii. Tap Modified to sort the files by name, date, type or size.

iv. Tap a file to open it.

How to Delete files.

i. Go to the Files app.

ii. Select and hold a file.

iii. Select "Delete."

iv. Select "OK."

How to Share a file

i. Select and hold the file.

ii. Select the Share icon

How to print, add to Google drive or perform other actions

i. Select a file to open.

ii. Go to the top right

iii. Tap More to get more options.

How to download videos to your Android phone

i. Ensure your device is connected to a working Wi-Fi or Mobile network.

ii. Go to the Google Play Movies & TV app.

iii. Select Library.

iv. Tap the Download icon next to the movie or TV episode you want to download.

Tap download icon again to stop a download that is in progress.

Enhanced Gaming

With the introduction of Android 10, mobile gamers now have access to some necessary additional functionality allowing you to connect your various Bluetooth gamepads easily. The Pixel 4 comes with out-of-the-box support for PS4, Xbox One, and even the Nintendo Switch Pro controllers.

How to connect a gamepad to Pixel 4
1. Power on your gamepad.
2. Go to Settings.
3. Tap Connected devices.
4. Tap Pair new devices.
5. Select your gamepad.

You will now be able to use the gamepad you like with no issues, once syncing and the connection is complete. However, note that not all games support a controller as standard.

Quick Vibrate mode

- This feature provides you with a quick way of toggling your Pixel 4 to the vibrate mode without having to unlock your device.
- To activate the mode, tap the power button and volume up button simultaneously and you should feel your phone has it vibrates. This indicates you have successfully switched to the vibrate mode.

Continuously identify songs around you with Now Playing

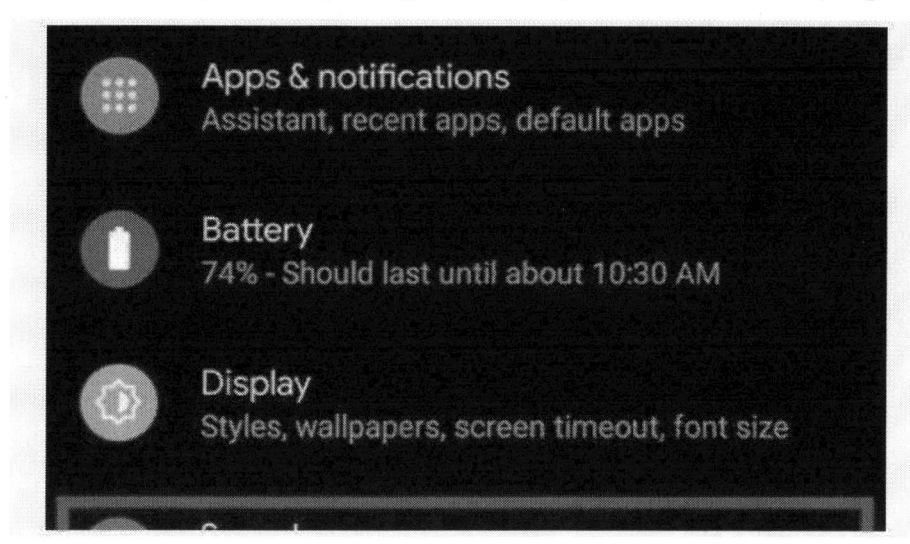

Apps & notifications
Assistant, recent apps, default apps

Battery
74% - Should last until about 10:30 AM

Display
Styles, wallpapers, screen timeout, font size

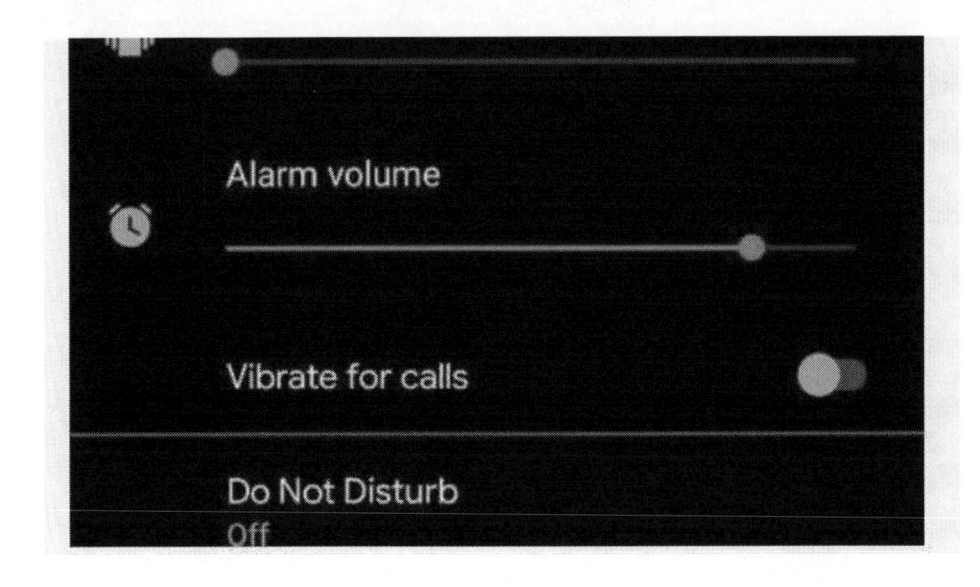

The Now Playing feature can now detect and identify music around you ensuring they are neatly added to the chronologically sorted archive for you to access later. It will not consume much of your battery because it doesn't require the internet to function.

However, note that it will not identify every song everywhere you go. The feature allows you to identify songs discretely, send them to the master log or display the name of songs on the lock screen.

Take the following steps to see songs on your lock screen:

1. Go to Settings.

2. Tap Sound.

3. Tap Now Playing.

4. Toggle on Show on lock screen.

You can also access the master log under the Now Playing History on the screen.

Chapter 22: Android 10 Navigation Tips

The Android 10 has made a huge change when it comes to navigating on the Pixel phones. Google has embraced the use of gestures and has added more and more of these gestures thereby allowing users to use their device without having to use the home button.

Once you tap or swipe, the Android 10 device will indicate a small line indicating that you are about to interact with the device.

How to access Overview:

• You can access the overview by swiping slowly from the bottom to about 1 cm/half an inch thereby popping up a display with your recent apps appearing in cards on the left.

• You will be taken to the apps tray from where you can scroll through all your apps horizontally.

Go Back to the home screen

Return to the Home screen by swiping up from the bottom of the screen. You will be navigated to the wallpaper automatically.

How to Open the Apps Tray

This also involves swiping up from the bottom of the display but at a faster rate. Just like swiping to return to the home screen. In a situation where you are in an app and like to access the apps tray:

1. Swipe slowly to access the Overview.

2. Swipe up.

3. Double tapping on the recent apps button switches you between your current and previous app in the Android Oreo and later.

This feature is however replaced in Pie by just swiping across the bottom display. All you need to do is swipe from the left to right. This will enable you to skip between apps as much as you like.

Go Back

Without the back button, you can now take a swipe in from the right or left which will activate the back feature irrespective of your present screen.

Change the sensitivity of gesture navigation

You can change the sensitivity of the gestures by taking the following steps:

1. Go to Settings.

2. Select System.

3. Select Gestures.

4. Tap System navigation.

You will find the settings option available for the gesture navigation

Bring back the three-button navigation bar

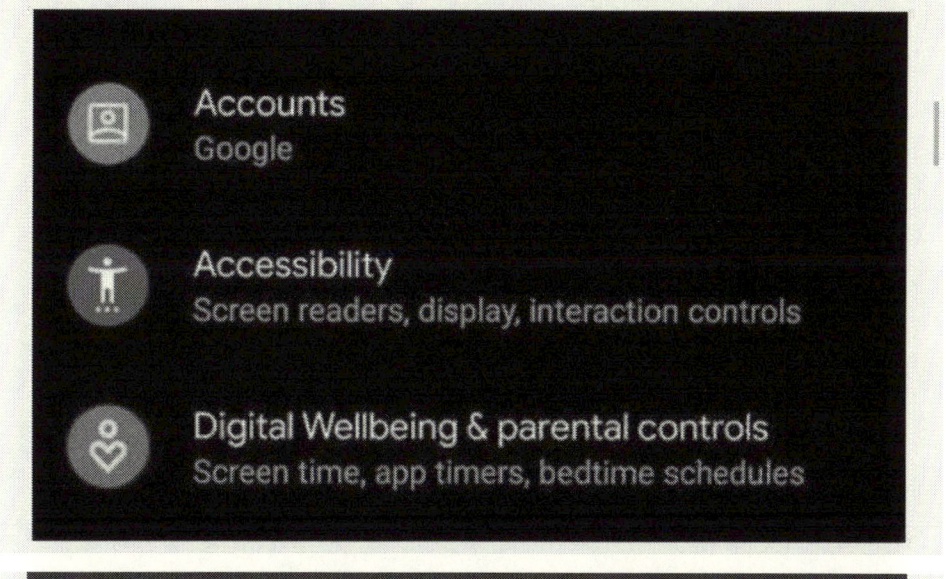

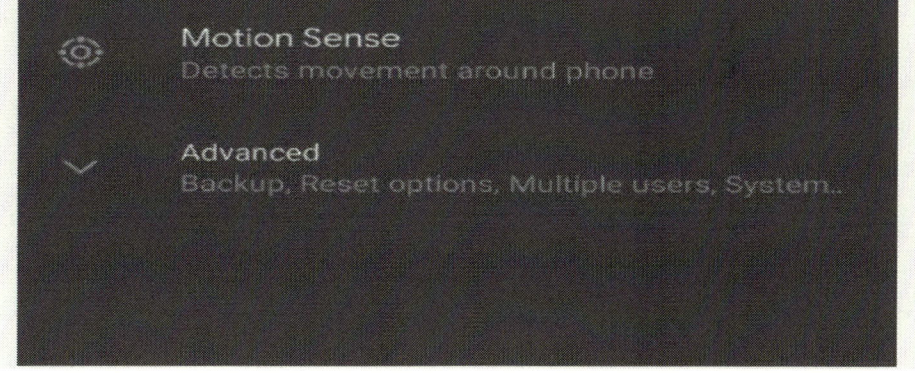

Google has officially switched over to the gesture-based navigation with the new Android 10 by default working similarly to the iPhone's gestures, though back function makes this feature unique in Android.

There are slide-over menus still made available currently which can easily be accessed by swiping back just like the same gesture that now presents itself as a back function. You can access this menu now by tapping on the left edge of the screen and holding for a bit till you can get it pulled to the right where you will access the menu underneath.

These menus can also be accessed from a hamburger menu icon located at the corner of your app's interface, so you can decide not to use the gesture if you don't want to

1. Go to Settings.
2. Select System.
3. Select Gestures.
4. Tap on System navigation.
5. Select between the three-button navigation and gestures.

It should be noted that it might be difficult to get the Google Assistant to work during this moment.

Get Pop-Up/Floating Navigation

• To save you the stress of having to constantly switch apps, get the Google Maps to provide you with floating navigation.

• All you need to activate this feature is to start your navigation in Google Maps and then tap the home button.

This will make the Maps shrink to a floating live window that can be placed anywhere on the screen.

Close All Open Apps View Overview

Shut all apps down by either swiping them all away at once to the top in Overview, or by scrolling all the way to the end of the list. Then hit the "clear all" option.

Enable App Suggestions

The Android 10 is built with suggestions for the apps you might prefer. These suggestions are originated from your app use enabling it to suggest apps you might need to instantly access. You can find this option in:

1. The home settings.

2. Tap suggestions and choose if you want to have it enabled or not.

Customize the Style Of Your Interface

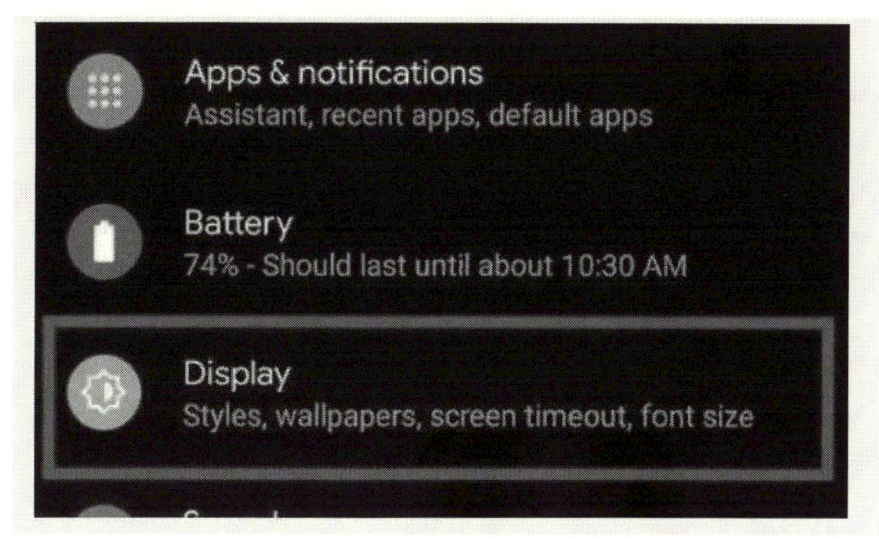

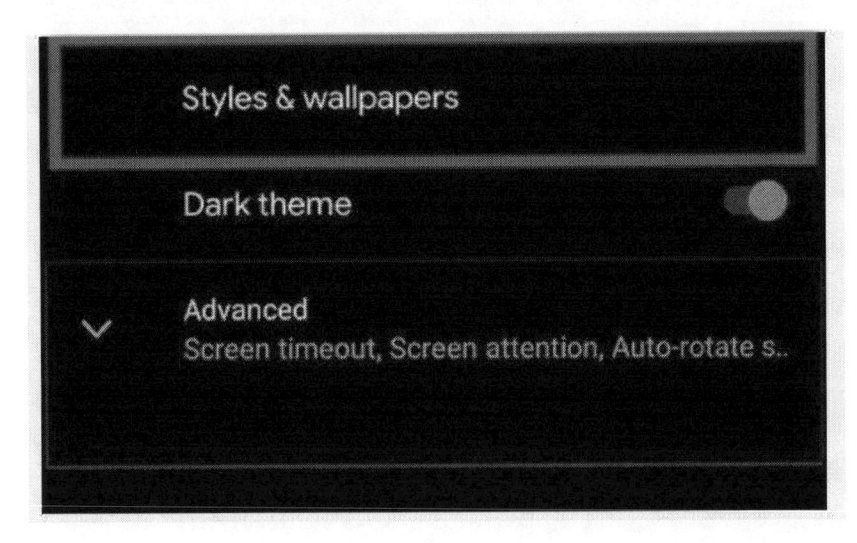

> **Styles & wallpapers**
>
> **Dark theme**
>
> ⌄ **Advanced**
> Screen timeout, Screen attention, Auto-rotate s..

- Google now provides you with the option of choosing from various kinds of accent colors for the user interface, fonts, and different icon shapes as well as different wallpapers.

- They also have a set of Pixel 4 themes, which can change things such as the wallpaper, shape, as well as the font of the app icons. You can create your styles manually if you don't like the one used as default.

Take the following steps to select your style:

1. Go to Settings.
2. Tap Display.
3. Select Styles and wallpapers.

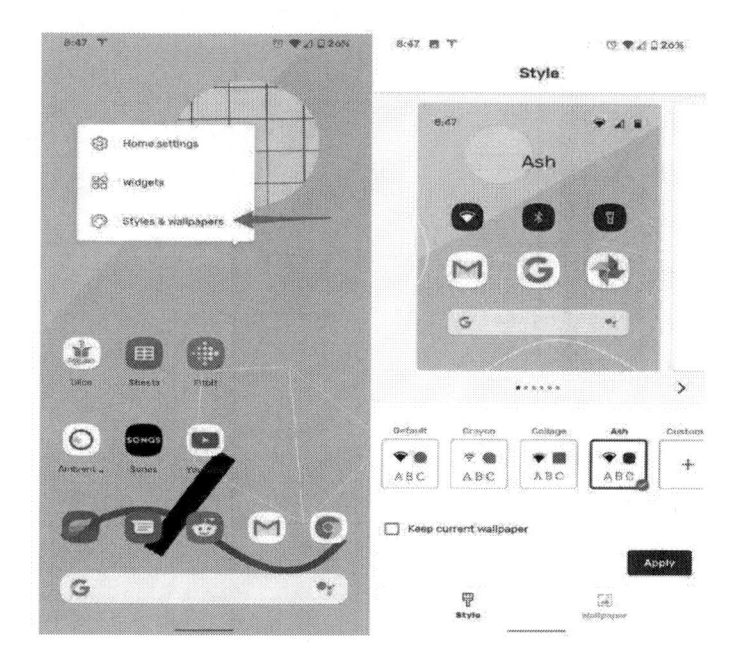

For easy access to the Styles section:

1. Make a long press on an empty space available on the Home screen.
2. Select Styles & wallpapers.
3. Go to the bottom of the screen and select the Styles tab.

You can scroll through different styles that are provided by Google or selecting the "**Custom**" option to create yours.

Turn On 90Hz All the Time

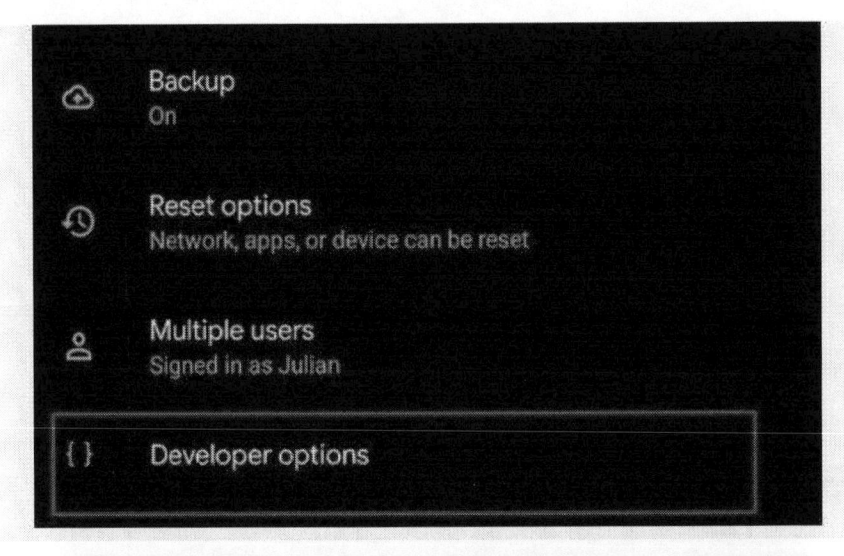

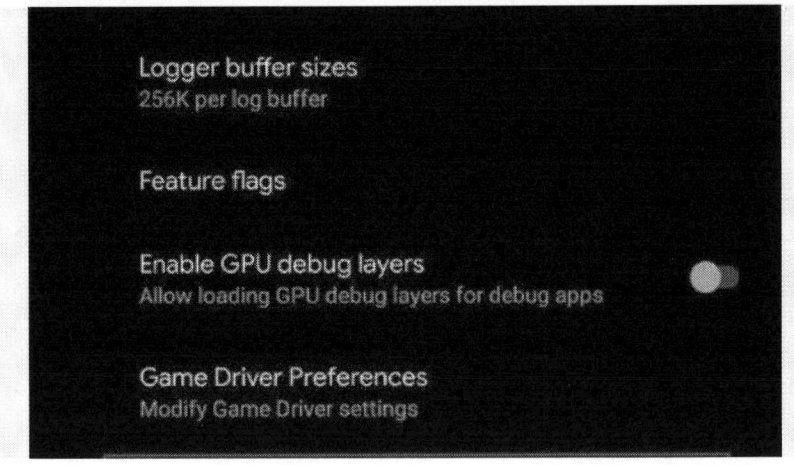

> Backup
> On
>
> Reset options
> Network, apps, or device can be reset
>
> Multiple users
> Signed in as Julian
>
> { } Developer options

> Logger buffer sizes
> 256K per log buffer
>
> Feature flags
>
> Enable GPU debug layers
> Allow loading GPU debug layers for debug apps
>
> Game Driver Preferences
> Modify Game Driver settings

Warning: This feature will have an enormous impact on the battery life of your device if turned on.

The screen of both the Pixel 4 and the Pixel 4 XL supports the 90Hz refresh rate which means they can deliver 90 Frames per second (FPS) unlike we have in most of the traditional phones that deliver 60FPS.

Using this feature makes playing games, scrolling through pages, and the interface use more responsive.

However, Google restricts the use of this feature due to its impact on the battery and as a result, it is not enabled all the time. It should be on by default but can check to be sure by taking the following steps:

1. Go to Settings.
2. Select Display.
3. Select Advanced.
4. Tap Smooth Display.

This feature can also be overridden by taking the following steps: First, turn on the Developers options.

1. Go to Settings.
2. Tap About phone.
3. Tap on Build number repeatedly until you are prompted with a note indicating you have turned on the Developer option.

Then go back and do the following.

1. Go to System.
2. Select Advanced.
3. Select Developer options.
4. Tap Force 90 Hz refresh rate and toggle it on.

However, remember that this will consume much of your battery life.

Digital Wellbeing

In case you are worried about always spending too much time on your phone, you are advised to go to the settings and search for the "Digital Wellbeing" option.

This feature also appears as an app. Though this option will not provide you with the details about your app and phone usage, you can use it to set timers and access other functions to always remind you to switch your phone off.

Using Wind down disconnect

Wind Down is a great way to proactively protect your bedtime. It is A motion that will slowly move your phone to greyscale and then switches it to the do not disturb. This will help you take your eyes off your phone.

All you need is going to the Digital Wellbeing to find and set the feature up.

Check for Android updates

If you like to get the latest version of the software, take the following steps:

1. Go to the Settings.

2. Tap System.

3. Tap Advanced.

4. Select System update.

You can manually check for any updates that have been pushed here. Finding the Android 10 Easter egg

1. Go to the Settings.
2. Tap About phone.
3. Select your Android version.
4. Tap Android 10 and it flips to the page that says Android 10 on it.
5. Double and hold the 1 and it rotates.
6. Drop to 0 and it creates a Q and the background will start to scroll.

Search Settings

- You can search the settings instead of having to go through it all. All you need to do is to open the Settings menu and you will be provided with a search bar at the top.
- This will help you to search for any settings on your phone easily.

Conclusion

If you have found this guide useful, please leave a review on Amazon.

Best regards!!

Index

A

Adaptive brightness, 154, 155
alarms, 76, 77, 83
Artificial intelligence, 128, 134
automatic unlock, 106, 107

B

backup, 14, 20, 22, 24, 27, 28, 29, 45, 175
Battery, 6, 152
Block, 73, 74, 77, 85, 86, 87, 93, 98

C

calendar, 43, 45, 81, 201, 202, 203
Calendar, 13, 45, 81, 200, 201

camera, 22, 94, 137, 138, 147, 177, 184, 185, 186, 187, 188, 189, 190, 192, 193
color, 99, 147, 155, 156, 169, 175, 201, 205
colors, 153, 165, 220
Contacts, 12, 13, 14, 30, 31, 33, 34, 35, 36, 37, 43, 59, 163
Control Notifications, 51

D

Dark Theme, 98
Developer Options, 145, 146
Digital Wellbeing, 224
Do Not Disturb, 52, 53, 54, 57, 58, 76, 77, 78, 79, 80, 81, 82, 83, 84, 85, 87, 88, 101

Drag, 49, 61, 88, 102, 136, 139, 140, 141, 159

E

Emergency, 90, 162, 163
Exporting, 31

F

Face Groups, 181
face unlock, 108, 109
Face Unlock, 108
folder, 42, 140
font size, 154, 161, 201, 202

G

Gestures, 49, 50, 54, 57, 128, 134, 142, 187, 216, 218
Gmail, 30, 31, 32, 33, 46, 199, 200, 207
Google Assistant, 53, 75, 120, 121, 122, 123, 124, 125, 126, 127, 128, 129, 130, 132, 133, 134, 135, 171, 218
Google Cloud, 204, 206, 210
Google drive, 43, 44, 45, 210, 211
Google Lens, 185, 192, 193

H

Home screen, 122, 135, 139, 141, 142, 169, 170, 171, 173, 203, 215, 221
Home Screen, 109, 135

I

IMEI, 15
Importing, 30, 32, 35
iPhone, 18, 19, 20, 21, 22, 25, 34, 43, 44, 45, 46, 47, 50, 175, 217
iTunes, 20, 39

L

lock screen, 47, 52, 58, 89, 101, 103, 104, 105, 108, 109, 151, 152, 162, 187, 214, 215
Lockdown, 113

M

make calls, 61, 62, 71
Make Calls, 61
messages, 12, 13, 20, 21, 23, 63, 68, 76, 79, 91, 93, 94, 96, 97, 100, 195, 199

Motion Sense, 54, 55, 110, 111, 134, 148, 170, 171
Moving Of Items, 48
multitasking, 167
music, 12, 13, 37, 38, 39, 40, 41, 42, 43, 47, 55, 83, 84, 89, 133, 134, 196, 214
Music, 12, 37, 38, 39, 40, 41, 42, 43, 47, 48

N

Networking, 4, 8
Night Sight, 191

O

Ok Google, 29, 120, 121, 122, 123, 124, 129, 158
Online Library, 41

P

phone by lifting, 51
Phone Call, 59, 60
Photos, 12, 13, 22, 23, 24, 25, 26, 27, 28, 43, 45, 46, 165, 174, 175, 176, 177, 178, 180, 181, 182, 184, 192
Pokemon, 134, 170, 171, 172

Print, 197, 198, 199, 200, 201, 202, 203, 204, 205, 210
Print Service, 198

Q

quick settings, 157, 159, 169

R

Reach to check, 148
refresh rate, 145, 148, 149, 222, 223

S

Safety, 138
screen saver, 164, 165, 166
Screenshot, 173, 174
Search Settings, 225
shortcuts, 1, 135, 136, 137, 138, 140, 186
Silencing sounds, 86
SIM card, 12, 15, 18, 21, 30, 35, 36, 46
Smart Displays, 130
Smart Lock, 106, 114, 115, 116, 118, 119
Spam, 72, 73, 74

228

T

The Do Not Disturb, 52
transcribe, 65
Transfer of Data, 11, 17
TTY, 69, 70

U

Unblock A Phone Number, 74
Use TTY or RTT with calls, 69

V

videos, 12, 13, 22, 23, 24, 25, 26, 27, 28, 43, 44, 46, 47, 83, 84, 93, 96, 137, 169, 177, 184, 186, 188, 195, 196, 212
voicemail, 63, 64, 65, 66, 67, 73, 74
VPN, 4, 5, 6

W

Wallpaper, 169, 170, 172
Wi-Fi, 2, 3, 4, 5, 6, 7, 8, 9, 10, 16, 24, 27, 28, 29, 40, 44, 62, 63, 112, 157, 158, 197, 212
Windows, 37, 42

Made in the USA
Monee, IL
10 February 2020